KICKING Logbook

Shawn Kovacich

AUTHOR APPROVED SHOOT/NO SHOOT EDITION

Printed in the United States of America

Library of Congress Control Number: 2019912454

ISBN #978-1-7336809-2-9

Table of Contents

Logbook Entry (continued)

Disclaimer

Please note that the author, Shawn Kovacich, and the publisher, **Shoot-NO-Shoot Productions**, Chikara Kan, Inc., of this logbook are **Not Responsible** in any manner whatsoever, for any injury, which may occur by reading and/or following the instructions located within this logbook. The distributor and any and all other individuals and or businesses connected with the production and/or distribution of this book also disclaim any and all liability for any damage or injury of any kind that may result from the appropriate use or inappropriate misuse of any information contained within this book.

Due to the fact that the techniques you will be executing are physical in nature, it is essential that the reader or readers of this book consult a qualified and competent physician before executing any physical techniques while keeping track of your progress in this logbook.

This book and all the information contained therein is intended to be used as a supplemental training aid, and should be used as such, under the guidance of a qualified and competent martial arts/self-defense instructor.

This book and the information contained within are for informational purposes only!

Copyrights & Trademarks

SAFETY

Due to the physical nature of the activities that you will be participating in while utilizing this logbook, it is highly recommended that you first consult a qualified and competent physician before starting any physical training routine. Regardless of your current physical condition. Remember, an ounce of prevention is worth a pound of cure. And also remember, **Safety First, Foremost, and Always!**

AND FINALLY...

You would do well to remember that progress should be measured in inches, not in miles. You are not going to execute 300 or 400 repetitions of a particular kick and then all of a sudden be showing a remarkable improvement in your kicking skills. It takes time, proper practice, full immersion and active participation, and consistency in practice in order to achieve positive results. You probably won't see any big changes after just a few hundred repetitions, but you should be seeing some significant improvement after the first 1,000 repetitions. And, you should continue to show improvement after every 1,000 repetitions of a particular kick that you execute, if you have fully immersed yourself both physically and mentally, while actively participating in each and every training session.

In addition to that, if you have also been studying and refining your kicks under the watchful eye of a qualified and competent martial arts instructor, then you will make even greater progress in honing your kicking skills.

HOW TO USE THIS LOGBOOK

In order to obtain the full benefits of this **Kicking Logbook**, you will want to follow these simple instructions on how to best utilize this logbook. This logbook should be used along with your active participation with the various kicks that you are currently practicing. These kicks that you have learned and/or are learning, should be practiced on a regular and consistent basis while keeping an extensive and detailed record of the number of repetitions that you are executing with each kick.

However, please make sure that you are first doing the kicks correctly before you start keeping track of the thousands of repetitions that you are going to start keeping track of. Because if you are doing the kicks incorrectly the first time, and you continue to practice them incorrectly, how do you think you are going to execute them the 6,000th time?

COVER & SPINE

When you look at the front cover and the spine, you will see a small white box in the lower left hand corner on the front cover and a smaller white box on the spine. These are include in order to give you a space in which to write down the volume number or other such indicator of the logbook that you are currently using.

LOGBOOK ENTRY (on the **"Table of Contents"** pages)

This is where you will write down the name of the **"Kick"** that you are practicing.

KICK

Write down the name of the **"Kick"** that you are going to be practicing and keeping meticulous track of. For example; Roundhouse Kick, Side Kick, Jump Turning Back Kick, etc.

START DATE

This is where you will enter the **"Date"** that you started keeping track of the number of repetitions that you executed of this particular kick. Be sure and include the month, day, and year.

KICKING LEG

Are you keeping track of the number of repetitions of this particular kick with just the **"Left"** leg, **"Right"** leg, or both legs (**"Alt"**)? If you are keeping track of the number of repetitions executed with both legs, make sure that you have executed 10 repetitions of this particular kick with both the right and left leg before circling the number.

KICKING HEIGHT

Are you executing each repetition of this particular kick at a **"Low"** section level, **"Mid"** section level, or a **"High"** section level? If you are executing the same kick at different levels, I would recommend using a separate section for each kick done at each height. For example; Roundhouse Low Section, Roundhouse Mid section, and Roundhouse High section, should each have a separate page in which you are keeping track of those repetitions executed.

NOTES

This section allows you to write down any particular **"Notes"** that you may have concerning this particular kick that you are keeping track of.

SLOW REPS

This section allows you to keep track of the number of slow repetitions or **"Slow Reps"** that you execute throughout the time frame that you are also executing your regular repetitions of this particular kick. My personal recommendation is that you should execute at least 10 **"Slow Reps"** at the beginning and at the end of every session that you are working on this particular kick.

For example if you are going to perform 100 **Roundhouse Kicks** alternating **"Alt"** legs at a **"High"** section level by executing 50 kicks with each leg. You should start out be executing 10 slow repetitions with each leg followed by 50 repetitions with each leg at normal speed and then finish up with another 10 slow repetitions with each leg. That would give you a total of 140 kicks executed in that particular training session, 20 **"Slow Reps"** with each leg and 50 **"Reg Reps"** with each leg.

REG REPS

The regular repetitions or **"Reg Reps"** section allows you to circle or check off the number of regular repetitions of this particular kick that you are keeping track of in increments of 10. So, for every 10 repetitions that you execute of a particular kick, you would check off the next 10 repetition increment. For example; if you are just starting out and you just finished executing say 50 repetitions of **Roundhouse Kick** with your **"Left"** leg at a **"High"** section level, you would then circle or check off the following numbers; 10, 20, 30, 40, and 50.

However, if you are executing kicks alternating **"Alt"** with both legs, you would need to make sure that you are executing an equal number of kicks with each leg that when divided by 2 would give you a number ending in zero. For example; if you executed a total of 70 kicks (35 on each leg) that wouldn't be divisible by two in order to come up with a number ending in zero. So you would need to execute 5 more kicks with each leg in order to come up with a total of 80 kicks which you would then divide by 2 in order to come up with a number ending in zero, which would be 40 kicks executed with each leg. You would then circle or check off the following numbers (if you started out at zero); 10, 20, 30, and 40.

NOTES

This **"Notes"** section, allows you the opportunity to write down any particular notes that you may have concerning this particular kick after every 1,000 repetitions executed. There is no notes section after the **"Slow Reps,"** as your slow repetitions are intended to be executed at various stages during the execution of the 5,000 regular repetitions or **"Reg Reps."** Be sure and include the month, day, and year. Use the **"Additional Notes"** section included at the end of this logbook to keep track of any additional items that you feel you need to keep track of during the course of utilizing this logbook.

END DATE

This is where you will enter the **"Date"** in which you completed the 1,000 slow repetitions **"Slow Reps"** and the 5,000 regular repetitions **"Reg Reps"** of this particular kick.

CODE KEY

This section was included in order to allow you the opportunity to personalize the way you keep track of your progress by allowing you to use a variety of symbols, ink colors, etc., in order to keep track of any particularities that you wanted to include in your own personal way to add to your training regiment. I have included some various possibilities that you could use, or come up with your own that may fit your own particular needs. So without further ado, let's get started.

BLACK INK = Physically and mentally an average training routine.

BLUE INK = Physically and mentally not feeling well. Not a typical training routine.

RED INK = Really pumped up both physically and mentally. Outstanding training routine.

X (over the number) = Executed that particular number of repetitions while doing line drills, etc.

CIRCLE (around the number) = Executed that particular number of repetitions while kicking in the air.

SQUARE (around the number) = Executed that particular number of repetitions while striking an object, ie; heavy bag, kicking "Shield," etc.

TRIANGLE (around the number) = Executed that particular number of repetitions while using a training partner as a target.

CODE KEY

Here is the **"Code Key"** for you to personalize to your own particular needs and training regiment. I would strongly recommend that you read through the **"How To Use This Logbook"** section prior to using this logbook in order to maximize its use to its fullest potential. Be sure and fill this out prior to starting your logbook.

BLACK INK ______________________________

____

BLUE INK ______________________________

____

RED INK ______________________________

____

X ______________________________

____

CIRCLE ______________________________

____

SQUARE ______________________________

____

TRIANGLE ______________________________

____

Kick:____**Start Date:** __________

Kicking Leg: Left - Right - Alt. **Kicking Height: Low - Mid - High**

Notes:____

____

____

Slow Reps: 10 - 20 - 30 - 40 - 50 - 60 - 70 - 80 - 90 - 100 - 110 - 120 - 130
140 - 150 - 160 - 170 - 180 - 190 - 200 - 210 - 220 - 230 - 240 - 250 - 260
270 - 280 - 290 - 300 - 310 - 320 - 330 - 340 - 350 - 360 - 370 - 380 - 390
400 - 410 - 420 - 430 - 440 - 450 - 460 - 470 - 480 - 490 - 500 - 510 - 520
530 - 540 - 550 - 560 - 570 - 580 - 590 - 600 - 610 - 620 - 630 - 640 - 650
660 - 670 - 680 - 690 - 700 - 710 - 720 - 730 - 740 - 750 - 760 - 770 - 780
790 - 800 - 810 - 820 - 830 - 840 - 850 - 860 - 870 - 880 - 890 - 900 - 910
920 - 930 - 940 - 950 - 960 - 970 - 980 - 990 - 1000

Reg Reps: 10 - 20 - 30 - 40 - 50 - 60 - 70 - 80 - 90 - 100 - 110 - 120 - 130
140 - 150 - 160 - 170 - 180 - 190 - 200 - 210 - 220 - 230 - 240 - 250 - 260
270 - 280 - 290 - 300 - 310 - 320 - 330 - 340 - 350 - 360 - 370 - 380 - 390
400 - 410 - 420 - 430 - 440 - 450 - 460 - 470 - 480 - 490 - 500 - 510 - 520
530 - 540 - 550 - 560 - 570 - 580 - 590 - 600 - 610 - 620 - 630 - 640 - 650
660 - 670 - 680 - 690 - 700 - 710 - 720 - 730 - 740 - 750 - 760 - 770 - 780
790 - 800 - 810 - 820 - 830 - 840 - 850 - 860 - 870 - 880 - 890 - 900 - 910
920 - 930 - 940 - 950 - 960 - 970 - 980 - 990 - 1000

Notes: ____

____

____

____

Reg Reps: 10 - 20 - 30 - 40 - 50 - 60 - 70 - 80 - 90 - 100 - 110 - 120 - 130
140 - 150 - 160 - 170 - 180 - 190 - 200 - 210 - 220 - 230 - 240 - 250 - 260
270 - 280 - 290 - 300 - 310 - 320 - 330 - 340 - 350 - 360 - 370 - 380 - 390
400 - 410 - 420 - 430 - 440 - 450 - 460 - 470 - 480 - 490 - 500 - 510 - 520
530 - 540 - 550 - 560 - 570 - 580 - 590 - 600 - 610 - 620 - 630 - 640 - 650
660 - 670 - 680 - 690 - 700 - 710 - 720 - 730 - 740 - 750 - 760 - 770 - 780
790 - 800 - 810 - 820 - 830 - 840 - 850 - 860 - 870 - 880 - 890 - 900 - 910
920 - 930 - 940 - 950 - 960 - 970 - 980 - 990 - 1000

Notes: ____

____

____

____

Reg Reps: 10 - 20 - 30 - 40 - 50 - 60 - 70 - 80 - 90 - 100 - 110 - 120 - 130
140 - 150 - 160 - 170 - 180 - 190 - 200 - 210 - 220 - 230 - 240 - 250 - 260
270 - 280 - 290 - 300 - 310 - 320 - 330 - 340 - 350 - 360 - 370 - 380 - 390
400 - 410 - 420 - 430 - 440 - 450 - 460 - 470 - 480 - 490 - 500 - 510 - 520
530 - 540 - 550 - 560 - 570 - 580 - 590 - 600 - 610 - 620 - 630 - 640 - 650
660 - 670 - 680 - 690 - 700 - 710 - 720 - 730 - 740 - 750 - 760 - 770 - 780
790 - 800 - 810 - 820 - 830 - 840 - 850 - 860 - 870 - 880 - 890 - 900 - 910
920 - 930 - 940 - 950 - 960 - 970 - 980 - 990 - 1000

Notes: ____

____

____

____

Reg Reps: 10 - 20 - 30 - 40 - 50 - 60 - 70 - 80 - 90 - 100 - 110 - 120 - 130
140 - 150 - 160 - 170 - 180 - 190 - 200 - 210 - 220 - 230 - 240 - 250 - 260
270 - 280 - 290 - 300 - 310 - 320 - 330 - 340 - 350 - 360 - 370 - 380 - 390
400 - 410 - 420 - 430 - 440 - 450 - 460 - 470 - 480 - 490 - 500 - 510 - 520
530 - 540 - 550 - 560 - 570 - 580 - 590 - 600 - 610 - 620 - 630 - 640 - 650
660 - 670 - 680 - 690 - 700 - 710 - 720 - 730 - 740 - 750 - 760 - 770 - 780
790 - 800 - 810 - 820 - 830 - 840 - 850 - 860 - 870 - 880 - 890 - 900 - 910
920 - 930 - 940 - 950 - 960 - 970 - 980 - 990 - 1000

Notes: ____

____

____

____

Reg Reps: 10 - 20 - 30 - 40 - 50 - 60 - 70 - 80 - 90 - 100 - 110 - 120 - 130
140 - 150 - 160 - 170 - 180 - 190 - 200 - 210 - 220 - 230 - 240 - 250 - 260
270 - 280 - 290 - 300 - 310 - 320 - 330 - 340 - 350 - 360 - 370 - 380 - 390
400 - 410 - 420 - 430 - 440 - 450 - 460 - 470 - 480 - 490 - 500 - 510 - 520
530 - 540 - 550 - 560 - 570 - 580 - 590 - 600 - 610 - 620 - 630 - 640 - 650
660 - 670 - 680 - 690 - 700 - 710 - 720 - 730 - 740 - 750 - 760 - 770 - 780
790 - 800 - 810 - 820 - 830 - 840 - 850 - 860 - 870 - 880 - 890 - 900 - 910
920 - 930 - 940 - 950 - 960 - 970 - 980 - 990 - 1000

Notes: ____

____

____

______________________________ **End Date:** ____________

Kick:________________________________ **Start Date:** _________

Kicking Leg: Left - Right - Alt. **Kicking Height: Low - Mid - High**

Notes:____

Slow Reps: 10 - 20 - 30 - 40 - 50 - 60 - 70 - 80 - 90 - 100 - 110 - 120 - 130
140 - 150 - 160 - 170 - 180 - 190 - 200 - 210 - 220 - 230 - 240 - 250 - 260
270 - 280 - 290 - 300 - 310 - 320 - 330 - 340 - 350 - 360 - 370 - 380 - 390
400 - 410 - 420 - 430 - 440 - 450 - 460 - 470 - 480 - 490 - 500 - 510 - 520
530 - 540 - 550 - 560 - 570 - 580 - 590 - 600 - 610 - 620 - 630 - 640 - 650
660 - 670 - 680 - 690 - 700 - 710 - 720 - 730 - 740 - 750 - 760 - 770 - 780
790 - 800 - 810 - 820 - 830 - 840 - 850 - 860 - 870 - 880 - 890 - 900 - 910
920 - 930 - 940 - 950 - 960 - 970 - 980 - 990 - 1000

Reg Reps: 10 - 20 - 30 - 40 - 50 - 60 - 70 - 80 - 90 - 100 - 110 - 120 - 130
140 - 150 - 160 - 170 - 180 - 190 - 200 - 210 - 220 - 230 - 240 - 250 - 260
270 - 280 - 290 - 300 - 310 - 320 - 330 - 340 - 350 - 360 - 370 - 380 - 390
400 - 410 - 420 - 430 - 440 - 450 - 460 - 470 - 480 - 490 - 500 - 510 - 520
530 - 540 - 550 - 560 - 570 - 580 - 590 - 600 - 610 - 620 - 630 - 640 - 650
660 - 670 - 680 - 690 - 700 - 710 - 720 - 730 - 740 - 750 - 760 - 770 - 780
790 - 800 - 810 - 820 - 830 - 840 - 850 - 860 - 870 - 880 - 890 - 900 - 910
920 - 930 - 940 - 950 - 960 - 970 - 980 - 990 - 1000

Notes: ____

Reg Reps: 10 - 20 - 30 - 40 - 50 - 60 - 70 - 80 - 90 - 100 - 110 - 120 - 130
140 - 150 - 160 - 170 - 180 - 190 - 200 - 210 - 220 - 230 - 240 - 250 - 260
270 - 280 - 290 - 300 - 310 - 320 - 330 - 340 - 350 - 360 - 370 - 380 - 390
400 - 410 - 420 - 430 - 440 - 450 - 460 - 470 - 480 - 490 - 500 - 510 - 520
530 - 540 - 550 - 560 - 570 - 580 - 590 - 600 - 610 - 620 - 630 - 640 - 650
660 - 670 - 680 - 690 - 700 - 710 - 720 - 730 - 740 - 750 - 760 - 770 - 780
790 - 800 - 810 - 820 - 830 - 840 - 850 - 860 - 870 - 880 - 890 - 900 - 910
920 - 930 - 940 - 950 - 960 - 970 - 980 - 990 - 1000

Notes: ____

Reg Reps: 10 - 20 - 30 - 40 - 50 - 60 - 70 - 80 - 90 - 100 - 110 - 120 - 130
140 - 150 - 160 - 170 - 180 - 190 - 200 - 210 - 220 - 230 - 240 - 250 - 260
270 - 280 - 290 - 300 - 310 - 320 - 330 - 340 - 350 - 360 - 370 - 380 - 390
400 - 410 - 420 - 430 - 440 - 450 - 460 - 470 - 480 - 490 - 500 - 510 - 520
530 - 540 - 550 - 560 - 570 - 580 - 590 - 600 - 610 - 620 - 630 - 640 - 650
660 - 670 - 680 - 690 - 700 - 710 - 720 - 730 - 740 - 750 - 760 - 770 - 780
790 - 800 - 810 - 820 - 830 - 840 - 850 - 860 - 870 - 880 - 890 - 900 - 910
920 - 930 - 940 - 950 - 960 - 970 - 980 - 990 - 1000

Notes: ____

____

____

____

Reg Reps: 10 - 20 - 30 - 40 - 50 - 60 - 70 - 80 - 90 - 100 - 110 - 120 - 130
140 - 150 - 160 - 170 - 180 - 190 - 200 - 210 - 220 - 230 - 240 - 250 - 260
270 - 280 - 290 - 300 - 310 - 320 - 330 - 340 - 350 - 360 - 370 - 380 - 390
400 - 410 - 420 - 430 - 440 - 450 - 460 - 470 - 480 - 490 - 500 - 510 - 520
530 - 540 - 550 - 560 - 570 - 580 - 590 - 600 - 610 - 620 - 630 - 640 - 650
660 - 670 - 680 - 690 - 700 - 710 - 720 - 730 - 740 - 750 - 760 - 770 - 780
790 - 800 - 810 - 820 - 830 - 840 - 850 - 860 - 870 - 880 - 890 - 900 - 910
920 - 930 - 940 - 950 - 960 - 970 - 980 - 990 - 1000

Notes: ____

____

____

____

Reg Reps: 10 - 20 - 30 - 40 - 50 - 60 - 70 - 80 - 90 - 100 - 110 - 120 - 130
140 - 150 - 160 - 170 - 180 - 190 - 200 - 210 - 220 - 230 - 240 - 250 - 260
270 - 280 - 290 - 300 - 310 - 320 - 330 - 340 - 350 - 360 - 370 - 380 - 390
400 - 410 - 420 - 430 - 440 - 450 - 460 - 470 - 480 - 490 - 500 - 510 - 520
530 - 540 - 550 - 560 - 570 - 580 - 590 - 600 - 610 - 620 - 630 - 640 - 650
660 - 670 - 680 - 690 - 700 - 710 - 720 - 730 - 740 - 750 - 760 - 770 - 780
790 - 800 - 810 - 820 - 830 - 840 - 850 - 860 - 870 - 880 - 890 - 900 - 910
920 - 930 - 940 - 950 - 960 - 970 - 980 - 990 - 1000

Notes: ____

____

____

______________________________ **End Date:** ____________

Kick: ______________________________ **Start Date:** __________

Kicking Leg: Left - Right - Alt. **Kicking Height: Low - Mid - High**

Notes: ____

____

____

Slow Reps: 10 - 20 - 30 - 40 - 50 - 60 - 70 - 80 - 90 - 100 - 110 - 120 - 130
140 - 150 - 160 - 170 - 180 - 190 - 200 - 210 - 220 - 230 - 240 - 250 - 260
270 - 280 - 290 - 300 - 310 - 320 - 330 - 340 - 350 - 360 - 370 - 380 - 390
400 - 410 - 420 - 430 - 440 - 450 - 460 - 470 - 480 - 490 - 500 - 510 - 520
530 - 540 - 550 - 560 - 570 - 580 - 590 - 600 - 610 - 620 - 630 - 640 - 650
660 - 670 - 680 - 690 - 700 - 710 - 720 - 730 - 740 - 750 - 760 - 770 - 780
790 - 800 - 810 - 820 - 830 - 840 - 850 - 860 - 870 - 880 - 890 - 900 - 910
920 - 930 - 940 - 950 - 960 - 970 - 980 - 990 - 1000

Reg Reps: 10 - 20 - 30 - 40 - 50 - 60 - 70 - 80 - 90 - 100 - 110 - 120 - 130
140 - 150 - 160 - 170 - 180 - 190 - 200 - 210 - 220 - 230 - 240 - 250 - 260
270 - 280 - 290 - 300 - 310 - 320 - 330 - 340 - 350 - 360 - 370 - 380 - 390
400 - 410 - 420 - 430 - 440 - 450 - 460 - 470 - 480 - 490 - 500 - 510 - 520
530 - 540 - 550 - 560 - 570 - 580 - 590 - 600 - 610 - 620 - 630 - 640 - 650
660 - 670 - 680 - 690 - 700 - 710 - 720 - 730 - 740 - 750 - 760 - 770 - 780
790 - 800 - 810 - 820 - 830 - 840 - 850 - 860 - 870 - 880 - 890 - 900 - 910
920 - 930 - 940 - 950 - 960 - 970 - 980 - 990 - 1000

Notes: ____

____

____

____

Reg Reps: 10 - 20 - 30 - 40 - 50 - 60 - 70 - 80 - 90 - 100 - 110 - 120 - 130
140 - 150 - 160 - 170 - 180 - 190 - 200 - 210 - 220 - 230 - 240 - 250 - 260
270 - 280 - 290 - 300 - 310 - 320 - 330 - 340 - 350 - 360 - 370 - 380 - 390
400 - 410 - 420 - 430 - 440 - 450 - 460 - 470 - 480 - 490 - 500 - 510 - 520
530 - 540 - 550 - 560 - 570 - 580 - 590 - 600 - 610 - 620 - 630 - 640 - 650
660 - 670 - 680 - 690 - 700 - 710 - 720 - 730 - 740 - 750 - 760 - 770 - 780
790 - 800 - 810 - 820 - 830 - 840 - 850 - 860 - 870 - 880 - 890 - 900 - 910
920 - 930 - 940 - 950 - 960 - 970 - 980 - 990 - 1000

Notes: ____

____

____

____

Reg Reps: 10 - 20 - 30 - 40 - 50 - 60 - 70 - 80 - 90 - 100 - 110 - 120 - 130
140 - 150 - 160 - 170 - 180 - 190 - 200 - 210 - 220 - 230 - 240 - 250 - 260
270 - 280 - 290 - 300 - 310 - 320 - 330 - 340 - 350 - 360 - 370 - 380 - 390
400 - 410 - 420 - 430 - 440 - 450 - 460 - 470 - 480 - 490 - 500 - 510 - 520
530 - 540 - 550 - 560 - 570 - 580 - 590 - 600 - 610 - 620 - 630 - 640 - 650
660 - 670 - 680 - 690 - 700 - 710 - 720 - 730 - 740 - 750 - 760 - 770 - 780
790 - 800 - 810 - 820 - 830 - 840 - 850 - 860 - 870 - 880 - 890 - 900 - 910
920 - 930 - 940 - 950 - 960 - 970 - 980 - 990 - 1000

Notes: ____

____

____

____

Reg Reps: 10 - 20 - 30 - 40 - 50 - 60 - 70 - 80 - 90 - 100 - 110 - 120 - 130
140 - 150 - 160 - 170 - 180 - 190 - 200 - 210 - 220 - 230 - 240 - 250 - 260
270 - 280 - 290 - 300 - 310 - 320 - 330 - 340 - 350 - 360 - 370 - 380 - 390
400 - 410 - 420 - 430 - 440 - 450 - 460 - 470 - 480 - 490 - 500 - 510 - 520
530 - 540 - 550 - 560 - 570 - 580 - 590 - 600 - 610 - 620 - 630 - 640 - 650
660 - 670 - 680 - 690 - 700 - 710 - 720 - 730 - 740 - 750 - 760 - 770 - 780
790 - 800 - 810 - 820 - 830 - 840 - 850 - 860 - 870 - 880 - 890 - 900 - 910
920 - 930 - 940 - 950 - 960 - 970 - 980 - 990 - 1000

Notes: ____

____

____

____

Reg Reps: 10 - 20 - 30 - 40 - 50 - 60 - 70 - 80 - 90 - 100 - 110 - 120 - 130
140 - 150 - 160 - 170 - 180 - 190 - 200 - 210 - 220 - 230 - 240 - 250 - 260
270 - 280 - 290 - 300 - 310 - 320 - 330 - 340 - 350 - 360 - 370 - 380 - 390
400 - 410 - 420 - 430 - 440 - 450 - 460 - 470 - 480 - 490 - 500 - 510 - 520
530 - 540 - 550 - 560 - 570 - 580 - 590 - 600 - 610 - 620 - 630 - 640 - 650
660 - 670 - 680 - 690 - 700 - 710 - 720 - 730 - 740 - 750 - 760 - 770 - 780
790 - 800 - 810 - 820 - 830 - 840 - 850 - 860 - 870 - 880 - 890 - 900 - 910
920 - 930 - 940 - 950 - 960 - 970 - 980 - 990 - 1000

Notes: ____

____

____

______________________________ **End Date:** ___________

Kick:______________________________ **Start Date:** ________
Kicking Leg: Left - Right - Alt. **Kicking Height: Low - Mid - High**
Notes:____

____

____

Slow Reps: 10 - 20 - 30 - 40 - 50 - 60 - 70 - 80 - 90 - 100 - 110 - 120 - 130
140 - 150 - 160 - 170 - 180 - 190 - 200 - 210 - 220 - 230 - 240 - 250 - 260
270 - 280 - 290 - 300 - 310 - 320 - 330 - 340 - 350 - 360 - 370 - 380 - 390
400 - 410 - 420 - 430 - 440 - 450 - 460 - 470 - 480 - 490 - 500 - 510 - 520
530 - 540 - 550 - 560 - 570 - 580 - 590 - 600 - 610 - 620 - 630 - 640 - 650
660 - 670 - 680 - 690 - 700 - 710 - 720 - 730 - 740 - 750 - 760 - 770 - 780
790 - 800 - 810 - 820 - 830 - 840 - 850 - 860 - 870 - 880 - 890 - 900 - 910
920 - 930 - 940 - 950 - 960 - 970 - 980 - 990 - 1000

Reg Reps: 10 - 20 - 30 - 40 - 50 - 60 - 70 - 80 - 90 - 100 - 110 - 120 - 130
140 - 150 - 160 - 170 - 180 - 190 - 200 - 210 - 220 - 230 - 240 - 250 - 260
270 - 280 - 290 - 300 - 310 - 320 - 330 - 340 - 350 - 360 - 370 - 380 - 390
400 - 410 - 420 - 430 - 440 - 450 - 460 - 470 - 480 - 490 - 500 - 510 - 520
530 - 540 - 550 - 560 - 570 - 580 - 590 - 600 - 610 - 620 - 630 - 640 - 650
660 - 670 - 680 - 690 - 700 - 710 - 720 - 730 - 740 - 750 - 760 - 770 - 780
790 - 800 - 810 - 820 - 830 - 840 - 850 - 860 - 870 - 880 - 890 - 900 - 910
920 - 930 - 940 - 950 - 960 - 970 - 980 - 990 - 1000
Notes: ____

____

____

____

Reg Reps: 10 - 20 - 30 - 40 - 50 - 60 - 70 - 80 - 90 - 100 - 110 - 120 - 130
140 - 150 - 160 - 170 - 180 - 190 - 200 - 210 - 220 - 230 - 240 - 250 - 260
270 - 280 - 290 - 300 - 310 - 320 - 330 - 340 - 350 - 360 - 370 - 380 - 390
400 - 410 - 420 - 430 - 440 - 450 - 460 - 470 - 480 - 490 - 500 - 510 - 520
530 - 540 - 550 - 560 - 570 - 580 - 590 - 600 - 610 - 620 - 630 - 640 - 650
660 - 670 - 680 - 690 - 700 - 710 - 720 - 730 - 740 - 750 - 760 - 770 - 780
790 - 800 - 810 - 820 - 830 - 840 - 850 - 860 - 870 - 880 - 890 - 900 - 910
920 - 930 - 940 - 950 - 960 - 970 - 980 - 990 - 1000
Notes: ____

____

____

____

Reg Reps: 10 - 20 - 30 - 40 - 50 - 60 - 70 - 80 - 90 - 100 - 110 - 120 - 130
140 - 150 - 160 - 170 - 180 - 190 - 200 - 210 - 220 - 230 - 240 - 250 - 260
270 - 280 - 290 - 300 - 310 - 320 - 330 - 340 - 350 - 360 - 370 - 380 - 390
400 - 410 - 420 - 430 - 440 - 450 - 460 - 470 - 480 - 490 - 500 - 510 - 520
530 - 540 - 550 - 560 - 570 - 580 - 590 - 600 - 610 - 620 - 630 - 640 - 650
660 - 670 - 680 - 690 - 700 - 710 - 720 - 730 - 740 - 750 - 760 - 770 - 780
790 - 800 - 810 - 820 - 830 - 840 - 850 - 860 - 870 - 880 - 890 - 900 - 910
920 - 930 - 940 - 950 - 960 - 970 - 980 - 990 - 1000

Notes: ______________________________

____

____

____

Reg Reps: 10 - 20 - 30 - 40 - 50 - 60 - 70 - 80 - 90 - 100 - 110 - 120 - 130
140 - 150 - 160 - 170 - 180 - 190 - 200 - 210 - 220 - 230 - 240 - 250 - 260
270 - 280 - 290 - 300 - 310 - 320 - 330 - 340 - 350 - 360 - 370 - 380 - 390
400 - 410 - 420 - 430 - 440 - 450 - 460 - 470 - 480 - 490 - 500 - 510 - 520
530 - 540 - 550 - 560 - 570 - 580 - 590 - 600 - 610 - 620 - 630 - 640 - 650
660 - 670 - 680 - 690 - 700 - 710 - 720 - 730 - 740 - 750 - 760 - 770 - 780
790 - 800 - 810 - 820 - 830 - 840 - 850 - 860 - 870 - 880 - 890 - 900 - 910
920 - 930 - 940 - 950 - 960 - 970 - 980 - 990 - 1000

Notes: ______________________________

____

____

____

Reg Reps: 10 - 20 - 30 - 40 - 50 - 60 - 70 - 80 - 90 - 100 - 110 - 120 - 130
140 - 150 - 160 - 170 - 180 - 190 - 200 - 210 - 220 - 230 - 240 - 250 - 260
270 - 280 - 290 - 300 - 310 - 320 - 330 - 340 - 350 - 360 - 370 - 380 - 390
400 - 410 - 420 - 430 - 440 - 450 - 460 - 470 - 480 - 490 - 500 - 510 - 520
530 - 540 - 550 - 560 - 570 - 580 - 590 - 600 - 610 - 620 - 630 - 640 - 650
660 - 670 - 680 - 690 - 700 - 710 - 720 - 730 - 740 - 750 - 760 - 770 - 780
790 - 800 - 810 - 820 - 830 - 840 - 850 - 860 - 870 - 880 - 890 - 900 - 910
920 - 930 - 940 - 950 - 960 - 970 - 980 - 990 - 1000

Notes: ______________________________

____

____

______________________________ **End Date:** __________

Kick:________________________________ **Start Date:** _________

Kicking Leg: Left - Right - Alt. **Kicking Height: Low - Mid - High**

Notes:____

____

____

Slow Reps: 10 - 20 - 30 - 40 - 50 - 60 - 70 - 80 - 90 - 100 - 110 - 120 - 130
140 - 150 - 160 - 170 - 180 - 190 - 200 - 210 - 220 - 230 - 240 - 250 - 260
270 - 280 - 290 - 300 - 310 - 320 - 330 - 340 - 350 - 360 - 370 - 380 - 390
400 - 410 - 420 - 430 - 440 - 450 - 460 - 470 - 480 - 490 - 500 - 510 - 520
530 - 540 - 550 - 560 - 570 - 580 - 590 - 600 - 610 - 620 - 630 - 640 - 650
660 - 670 - 680 - 690 - 700 - 710 - 720 - 730 - 740 - 750 - 760 - 770 - 780
790 - 800 - 810 - 820 - 830 - 840 - 850 - 860 - 870 - 880 - 890 - 900 - 910
920 - 930 - 940 - 950 - 960 - 970 - 980 - 990 - 1000

Reg Reps: 10 - 20 - 30 - 40 - 50 - 60 - 70 - 80 - 90 - 100 - 110 - 120 - 130
140 - 150 - 160 - 170 - 180 - 190 - 200 - 210 - 220 - 230 - 240 - 250 - 260
270 - 280 - 290 - 300 - 310 - 320 - 330 - 340 - 350 - 360 - 370 - 380 - 390
400 - 410 - 420 - 430 - 440 - 450 - 460 - 470 - 480 - 490 - 500 - 510 - 520
530 - 540 - 550 - 560 - 570 - 580 - 590 - 600 - 610 - 620 - 630 - 640 - 650
660 - 670 - 680 - 690 - 700 - 710 - 720 - 730 - 740 - 750 - 760 - 770 - 780
790 - 800 - 810 - 820 - 830 - 840 - 850 - 860 - 870 - 880 - 890 - 900 - 910
920 - 930 - 940 - 950 - 960 - 970 - 980 - 990 - 1000

Notes: ____

____

____

____

Reg Reps: 10 - 20 - 30 - 40 - 50 - 60 - 70 - 80 - 90 - 100 - 110 - 120 - 130
140 - 150 - 160 - 170 - 180 - 190 - 200 - 210 - 220 - 230 - 240 - 250 - 260
270 - 280 - 290 - 300 - 310 - 320 - 330 - 340 - 350 - 360 - 370 - 380 - 390
400 - 410 - 420 - 430 - 440 - 450 - 460 - 470 - 480 - 490 - 500 - 510 - 520
530 - 540 - 550 - 560 - 570 - 580 - 590 - 600 - 610 - 620 - 630 - 640 - 650
660 - 670 - 680 - 690 - 700 - 710 - 720 - 730 - 740 - 750 - 760 - 770 - 780
790 - 800 - 810 - 820 - 830 - 840 - 850 - 860 - 870 - 880 - 890 - 900 - 910
920 - 930 - 940 - 950 - 960 - 970 - 980 - 990 - 1000

Notes: ____

____

____

____

Reg Reps: 10 - 20 - 30 - 40 - 50 - 60 - 70 - 80 - 90 - 100 - 110 - 120 - 130
140 - 150 - 160 - 170 - 180 - 190 - 200 - 210 - 220 - 230 - 240 - 250 - 260
270 - 280 - 290 - 300 - 310 - 320 - 330 - 340 - 350 - 360 - 370 - 380 - 390
400 - 410 - 420 - 430 - 440 - 450 - 460 - 470 - 480 - 490 - 500 - 510 - 520
530 - 540 - 550 - 560 - 570 - 580 - 590 - 600 - 610 - 620 - 630 - 640 - 650
660 - 670 - 680 - 690 - 700 - 710 - 720 - 730 - 740 - 750 - 760 - 770 - 780
790 - 800 - 810 - 820 - 830 - 840 - 850 - 860 - 870 - 880 - 890 - 900 - 910
920 - 930 - 940 - 950 - 960 - 970 - 980 - 990 - 1000

Notes: ____

____

____

____

Reg Reps: 10 - 20 - 30 - 40 - 50 - 60 - 70 - 80 - 90 - 100 - 110 - 120 - 130
140 - 150 - 160 - 170 - 180 - 190 - 200 - 210 - 220 - 230 - 240 - 250 - 260
270 - 280 - 290 - 300 - 310 - 320 - 330 - 340 - 350 - 360 - 370 - 380 - 390
400 - 410 - 420 - 430 - 440 - 450 - 460 - 470 - 480 - 490 - 500 - 510 - 520
530 - 540 - 550 - 560 - 570 - 580 - 590 - 600 - 610 - 620 - 630 - 640 - 650
660 - 670 - 680 - 690 - 700 - 710 - 720 - 730 - 740 - 750 - 760 - 770 - 780
790 - 800 - 810 - 820 - 830 - 840 - 850 - 860 - 870 - 880 - 890 - 900 - 910
920 - 930 - 940 - 950 - 960 - 970 - 980 - 990 - 1000

Notes: ____

____

____

____

Reg Reps: 10 - 20 - 30 - 40 - 50 - 60 - 70 - 80 - 90 - 100 - 110 - 120 - 130
140 - 150 - 160 - 170 - 180 - 190 - 200 - 210 - 220 - 230 - 240 - 250 - 260
270 - 280 - 290 - 300 - 310 - 320 - 330 - 340 - 350 - 360 - 370 - 380 - 390
400 - 410 - 420 - 430 - 440 - 450 - 460 - 470 - 480 - 490 - 500 - 510 - 520
530 - 540 - 550 - 560 - 570 - 580 - 590 - 600 - 610 - 620 - 630 - 640 - 650
660 - 670 - 680 - 690 - 700 - 710 - 720 - 730 - 740 - 750 - 760 - 770 - 780
790 - 800 - 810 - 820 - 830 - 840 - 850 - 860 - 870 - 880 - 890 - 900 - 910
920 - 930 - 940 - 950 - 960 - 970 - 980 - 990 - 1000

Notes: ____

____

____

______________________________ **End Date:** ____________

Kick: ______________________________ **Start Date:** __________

Kicking Leg: Left - Right - Alt. **Kicking Height: Low - Mid - High**

Notes: ____

____

____

Slow Reps: 10 - 20 - 30 - 40 - 50 - 60 - 70 - 80 - 90 - 100 - 110 - 120 - 130
140 - 150 - 160 - 170 - 180 - 190 - 200 - 210 - 220 - 230 - 240 - 250 - 260
270 - 280 - 290 - 300 - 310 - 320 - 330 - 340 - 350 - 360 - 370 - 380 - 390
400 - 410 - 420 - 430 - 440 - 450 - 460 - 470 - 480 - 490 - 500 - 510 - 520
530 - 540 - 550 - 560 - 570 - 580 - 590 - 600 - 610 - 620 - 630 - 640 - 650
660 - 670 - 680 - 690 - 700 - 710 - 720 - 730 - 740 - 750 - 760 - 770 - 780
790 - 800 - 810 - 820 - 830 - 840 - 850 - 860 - 870 - 880 - 890 - 900 - 910
920 - 930 - 940 - 950 - 960 - 970 - 980 - 990 - 1000

Reg Reps: 10 - 20 - 30 - 40 - 50 - 60 - 70 - 80 - 90 - 100 - 110 - 120 - 130
140 - 150 - 160 - 170 - 180 - 190 - 200 - 210 - 220 - 230 - 240 - 250 - 260
270 - 280 - 290 - 300 - 310 - 320 - 330 - 340 - 350 - 360 - 370 - 380 - 390
400 - 410 - 420 - 430 - 440 - 450 - 460 - 470 - 480 - 490 - 500 - 510 - 520
530 - 540 - 550 - 560 - 570 - 580 - 590 - 600 - 610 - 620 - 630 - 640 - 650
660 - 670 - 680 - 690 - 700 - 710 - 720 - 730 - 740 - 750 - 760 - 770 - 780
790 - 800 - 810 - 820 - 830 - 840 - 850 - 860 - 870 - 880 - 890 - 900 - 910
920 - 930 - 940 - 950 - 960 - 970 - 980 - 990 - 1000

Notes: ____

____

____

____

Reg Reps: 10 - 20 - 30 - 40 - 50 - 60 - 70 - 80 - 90 - 100 - 110 - 120 - 130
140 - 150 - 160 - 170 - 180 - 190 - 200 - 210 - 220 - 230 - 240 - 250 - 260
270 - 280 - 290 - 300 - 310 - 320 - 330 - 340 - 350 - 360 - 370 - 380 - 390
400 - 410 - 420 - 430 - 440 - 450 - 460 - 470 - 480 - 490 - 500 - 510 - 520
530 - 540 - 550 - 560 - 570 - 580 - 590 - 600 - 610 - 620 - 630 - 640 - 650
660 - 670 - 680 - 690 - 700 - 710 - 720 - 730 - 740 - 750 - 760 - 770 - 780
790 - 800 - 810 - 820 - 830 - 840 - 850 - 860 - 870 - 880 - 890 - 900 - 910
920 - 930 - 940 - 950 - 960 - 970 - 980 - 990 - 1000

Notes: ____

____

____

____

Reg Reps: 10 - 20 - 30 - 40 - 50 - 60 - 70 - 80 - 90 - 100 - 110 - 120 - 130 140 - 150 - 160 - 170 - 180 - 190 - 200 - 210 - 220 - 230 - 240 - 250 - 260 270 - 280 - 290 - 300 - 310 - 320 - 330 - 340 - 350 - 360 - 370 - 380 - 390 400 - 410 - 420 - 430 - 440 - 450 - 460 - 470 - 480 - 490 - 500 - 510 - 520 530 - 540 - 550 - 560 - 570 - 580 - 590 - 600 - 610 - 620 - 630 - 640 - 650 660 - 670 - 680 - 690 - 700 - 710 - 720 - 730 - 740 - 750 - 760 - 770 - 780 790 - 800 - 810 - 820 - 830 - 840 - 850 - 860 - 870 - 880 - 890 - 900 - 910 920 - 930 - 940 - 950 - 960 - 970 - 980 - 990 - 1000

Notes: ____

____

____

____

Reg Reps: 10 - 20 - 30 - 40 - 50 - 60 - 70 - 80 - 90 - 100 - 110 - 120 - 130 140 - 150 - 160 - 170 - 180 - 190 - 200 - 210 - 220 - 230 - 240 - 250 - 260 270 - 280 - 290 - 300 - 310 - 320 - 330 - 340 - 350 - 360 - 370 - 380 - 390 400 - 410 - 420 - 430 - 440 - 450 - 460 - 470 - 480 - 490 - 500 - 510 - 520 530 - 540 - 550 - 560 - 570 - 580 - 590 - 600 - 610 - 620 - 630 - 640 - 650 660 - 670 - 680 - 690 - 700 - 710 - 720 - 730 - 740 - 750 - 760 - 770 - 780 790 - 800 - 810 - 820 - 830 - 840 - 850 - 860 - 870 - 880 - 890 - 900 - 910 920 - 930 - 940 - 950 - 960 - 970 - 980 - 990 - 1000

Notes: ____

____

____

____

Reg Reps: 10 - 20 - 30 - 40 - 50 - 60 - 70 - 80 - 90 - 100 - 110 - 120 - 130 140 - 150 - 160 - 170 - 180 - 190 - 200 - 210 - 220 - 230 - 240 - 250 - 260 270 - 280 - 290 - 300 - 310 - 320 - 330 - 340 - 350 - 360 - 370 - 380 - 390 400 - 410 - 420 - 430 - 440 - 450 - 460 - 470 - 480 - 490 - 500 - 510 - 520 530 - 540 - 550 - 560 - 570 - 580 - 590 - 600 - 610 - 620 - 630 - 640 - 650 660 - 670 - 680 - 690 - 700 - 710 - 720 - 730 - 740 - 750 - 760 - 770 - 780 790 - 800 - 810 - 820 - 830 - 840 - 850 - 860 - 870 - 880 - 890 - 900 - 910 920 - 930 - 940 - 950 - 960 - 970 - 980 - 990 - 1000

Notes: ____

____

____

______________________________ **End Date:** __________

Kick:________________________________ **Start Date:** ________

Kicking Leg: Left - Right - Alt. **Kicking Height: Low - Mid - High**

Notes:____

____

____

Slow Reps: 10 - 20 - 30 - 40 - 50 - 60 - 70 - 80 - 90 - 100 - 110 - 120 - 130
140 - 150 - 160 - 170 - 180 - 190 - 200 - 210 - 220 - 230 - 240 - 250 - 260
270 - 280 - 290 - 300 - 310 - 320 - 330 - 340 - 350 - 360 - 370 - 380 - 390
400 - 410 - 420 - 430 - 440 - 450 - 460 - 470 - 480 - 490 - 500 - 510 - 520
530 - 540 - 550 - 560 - 570 - 580 - 590 - 600 - 610 - 620 - 630 - 640 - 650
660 - 670 - 680 - 690 - 700 - 710 - 720 - 730 - 740 - 750 - 760 - 770 - 780
790 - 800 - 810 - 820 - 830 - 840 - 850 - 860 - 870 - 880 - 890 - 900 - 910
920 - 930 - 940 - 950 - 960 - 970 - 980 - 990 - 1000

Reg Reps: 10 - 20 - 30 - 40 - 50 - 60 - 70 - 80 - 90 - 100 - 110 - 120 - 130
140 - 150 - 160 - 170 - 180 - 190 - 200 - 210 - 220 - 230 - 240 - 250 - 260
270 - 280 - 290 - 300 - 310 - 320 - 330 - 340 - 350 - 360 - 370 - 380 - 390
400 - 410 - 420 - 430 - 440 - 450 - 460 - 470 - 480 - 490 - 500 - 510 - 520
530 - 540 - 550 - 560 - 570 - 580 - 590 - 600 - 610 - 620 - 630 - 640 - 650
660 - 670 - 680 - 690 - 700 - 710 - 720 - 730 - 740 - 750 - 760 - 770 - 780
790 - 800 - 810 - 820 - 830 - 840 - 850 - 860 - 870 - 880 - 890 - 900 - 910
920 - 930 - 940 - 950 - 960 - 970 - 980 - 990 - 1000

Notes: ____

____

____

____

Reg Reps: 10 - 20 - 30 - 40 - 50 - 60 - 70 - 80 - 90 - 100 - 110 - 120 - 130
140 - 150 - 160 - 170 - 180 - 190 - 200 - 210 - 220 - 230 - 240 - 250 - 260
270 - 280 - 290 - 300 - 310 - 320 - 330 - 340 - 350 - 360 - 370 - 380 - 390
400 - 410 - 420 - 430 - 440 - 450 - 460 - 470 - 480 - 490 - 500 - 510 - 520
530 - 540 - 550 - 560 - 570 - 580 - 590 - 600 - 610 - 620 - 630 - 640 - 650
660 - 670 - 680 - 690 - 700 - 710 - 720 - 730 - 740 - 750 - 760 - 770 - 780
790 - 800 - 810 - 820 - 830 - 840 - 850 - 860 - 870 - 880 - 890 - 900 - 910
920 - 930 - 940 - 950 - 960 - 970 - 980 - 990 - 1000

Notes: ____

____

____

____

Reg Reps: 10 - 20 - 30 - 40 - 50 - 60 - 70 - 80 - 90 - 100 - 110 - 120 - 130
140 - 150 - 160 - 170 - 180 - 190 - 200 - 210 - 220 - 230 - 240 - 250 - 260
270 - 280 - 290 - 300 - 310 - 320 - 330 - 340 - 350 - 360 - 370 - 380 - 390
400 - 410 - 420 - 430 - 440 - 450 - 460 - 470 - 480 - 490 - 500 - 510 - 520
530 - 540 - 550 - 560 - 570 - 580 - 590 - 600 - 610 - 620 - 630 - 640 - 650
660 - 670 - 680 - 690 - 700 - 710 - 720 - 730 - 740 - 750 - 760 - 770 - 780
790 - 800 - 810 - 820 - 830 - 840 - 850 - 860 - 870 - 880 - 890 - 900 - 910
920 - 930 - 940 - 950 - 960 - 970 - 980 - 990 - 1000

Notes: ____

____

____

____

Reg Reps: 10 - 20 - 30 - 40 - 50 - 60 - 70 - 80 - 90 - 100 - 110 - 120 - 130
140 - 150 - 160 - 170 - 180 - 190 - 200 - 210 - 220 - 230 - 240 - 250 - 260
270 - 280 - 290 - 300 - 310 - 320 - 330 - 340 - 350 - 360 - 370 - 380 - 390
400 - 410 - 420 - 430 - 440 - 450 - 460 - 470 - 480 - 490 - 500 - 510 - 520
530 - 540 - 550 - 560 - 570 - 580 - 590 - 600 - 610 - 620 - 630 - 640 - 650
660 - 670 - 680 - 690 - 700 - 710 - 720 - 730 - 740 - 750 - 760 - 770 - 780
790 - 800 - 810 - 820 - 830 - 840 - 850 - 860 - 870 - 880 - 890 - 900 - 910
920 - 930 - 940 - 950 - 960 - 970 - 980 - 990 - 1000

Notes: ____

____

____

____

Reg Reps: 10 - 20 - 30 - 40 - 50 - 60 - 70 - 80 - 90 - 100 - 110 - 120 - 130
140 - 150 - 160 - 170 - 180 - 190 - 200 - 210 - 220 - 230 - 240 - 250 - 260
270 - 280 - 290 - 300 - 310 - 320 - 330 - 340 - 350 - 360 - 370 - 380 - 390
400 - 410 - 420 - 430 - 440 - 450 - 460 - 470 - 480 - 490 - 500 - 510 - 520
530 - 540 - 550 - 560 - 570 - 580 - 590 - 600 - 610 - 620 - 630 - 640 - 650
660 - 670 - 680 - 690 - 700 - 710 - 720 - 730 - 740 - 750 - 760 - 770 - 780
790 - 800 - 810 - 820 - 830 - 840 - 850 - 860 - 870 - 880 - 890 - 900 - 910
920 - 930 - 940 - 950 - 960 - 970 - 980 - 990 - 1000

Notes: ____

____

____

_____________________________________ **End Date:** __________

Kick:________________________________ **Start Date:** ________
Kicking Leg: Left - Right - Alt. **Kicking Height: Low - Mid - High**
Notes:____

Slow Reps: 10 - 20 - 30 - 40 - 50 - 60 - 70 - 80 - 90 - 100 - 110 - 120 - 130
140 - 150 - 160 - 170 - 180 - 190 - 200 - 210 - 220 - 230 - 240 - 250 - 260
270 - 280 - 290 - 300 - 310 - 320 - 330 - 340 - 350 - 360 - 370 - 380 - 390
400 - 410 - 420 - 430 - 440 - 450 - 460 - 470 - 480 - 490 - 500 - 510 - 520
530 - 540 - 550 - 560 - 570 - 580 - 590 - 600 - 610 - 620 - 630 - 640 - 650
660 - 670 - 680 - 690 - 700 - 710 - 720 - 730 - 740 - 750 - 760 - 770 - 780
790 - 800 - 810 - 820 - 830 - 840 - 850 - 860 - 870 - 880 - 890 - 900 - 910
920 - 930 - 940 - 950 - 960 - 970 - 980 - 990 - 1000

Reg Reps: 10 - 20 - 30 - 40 - 50 - 60 - 70 - 80 - 90 - 100 - 110 - 120 - 130
140 - 150 - 160 - 170 - 180 - 190 - 200 - 210 - 220 - 230 - 240 - 250 - 260
270 - 280 - 290 - 300 - 310 - 320 - 330 - 340 - 350 - 360 - 370 - 380 - 390
400 - 410 - 420 - 430 - 440 - 450 - 460 - 470 - 480 - 490 - 500 - 510 - 520
530 - 540 - 550 - 560 - 570 - 580 - 590 - 600 - 610 - 620 - 630 - 640 - 650
660 - 670 - 680 - 690 - 700 - 710 - 720 - 730 - 740 - 750 - 760 - 770 - 780
790 - 800 - 810 - 820 - 830 - 840 - 850 - 860 - 870 - 880 - 890 - 900 - 910
920 - 930 - 940 - 950 - 960 - 970 - 980 - 990 - 1000
Notes: ____

Reg Reps: 10 - 20 - 30 - 40 - 50 - 60 - 70 - 80 - 90 - 100 - 110 - 120 - 130
140 - 150 - 160 - 170 - 180 - 190 - 200 - 210 - 220 - 230 - 240 - 250 - 260
270 - 280 - 290 - 300 - 310 - 320 - 330 - 340 - 350 - 360 - 370 - 380 - 390
400 - 410 - 420 - 430 - 440 - 450 - 460 - 470 - 480 - 490 - 500 - 510 - 520
530 - 540 - 550 - 560 - 570 - 580 - 590 - 600 - 610 - 620 - 630 - 640 - 650
660 - 670 - 680 - 690 - 700 - 710 - 720 - 730 - 740 - 750 - 760 - 770 - 780
790 - 800 - 810 - 820 - 830 - 840 - 850 - 860 - 870 - 880 - 890 - 900 - 910
920 - 930 - 940 - 950 - 960 - 970 - 980 - 990 - 1000
Notes: ____

Reg Reps: 10 - 20 - 30 - 40 - 50 - 60 - 70 - 80 - 90 - 100 - 110 - 120 - 130 - 140 - 150 - 160 - 170 - 180 - 190 - 200 - 210 - 220 - 230 - 240 - 250 - 260 - 270 - 280 - 290 - 300 - 310 - 320 - 330 - 340 - 350 - 360 - 370 - 380 - 390 - 400 - 410 - 420 - 430 - 440 - 450 - 460 - 470 - 480 - 490 - 500 - 510 - 520 - 530 - 540 - 550 - 560 - 570 - 580 - 590 - 600 - 610 - 620 - 630 - 640 - 650 - 660 - 670 - 680 - 690 - 700 - 710 - 720 - 730 - 740 - 750 - 760 - 770 - 780 - 790 - 800 - 810 - 820 - 830 - 840 - 850 - 860 - 870 - 880 - 890 - 900 - 910 - 920 - 930 - 940 - 950 - 960 - 970 - 980 - 990 - 1000

Notes: ______________________________

____

____

____

Reg Reps: 10 - 20 - 30 - 40 - 50 - 60 - 70 - 80 - 90 - 100 - 110 - 120 - 130 - 140 - 150 - 160 - 170 - 180 - 190 - 200 - 210 - 220 - 230 - 240 - 250 - 260 - 270 - 280 - 290 - 300 - 310 - 320 - 330 - 340 - 350 - 360 - 370 - 380 - 390 - 400 - 410 - 420 - 430 - 440 - 450 - 460 - 470 - 480 - 490 - 500 - 510 - 520 - 530 - 540 - 550 - 560 - 570 - 580 - 590 - 600 - 610 - 620 - 630 - 640 - 650 - 660 - 670 - 680 - 690 - 700 - 710 - 720 - 730 - 740 - 750 - 760 - 770 - 780 - 790 - 800 - 810 - 820 - 830 - 840 - 850 - 860 - 870 - 880 - 890 - 900 - 910 - 920 - 930 - 940 - 950 - 960 - 970 - 980 - 990 - 1000

Notes: ______________________________

____

____

____

Reg Reps: 10 - 20 - 30 - 40 - 50 - 60 - 70 - 80 - 90 - 100 - 110 - 120 - 130 - 140 - 150 - 160 - 170 - 180 - 190 - 200 - 210 - 220 - 230 - 240 - 250 - 260 - 270 - 280 - 290 - 300 - 310 - 320 - 330 - 340 - 350 - 360 - 370 - 380 - 390 - 400 - 410 - 420 - 430 - 440 - 450 - 460 - 470 - 480 - 490 - 500 - 510 - 520 - 530 - 540 - 550 - 560 - 570 - 580 - 590 - 600 - 610 - 620 - 630 - 640 - 650 - 660 - 670 - 680 - 690 - 700 - 710 - 720 - 730 - 740 - 750 - 760 - 770 - 780 - 790 - 800 - 810 - 820 - 830 - 840 - 850 - 860 - 870 - 880 - 890 - 900 - 910 - 920 - 930 - 940 - 950 - 960 - 970 - 980 - 990 - 1000

Notes: ______________________________

____

____

______________________________ **End Date:** ____________

Kick:________________________________ **Start Date:** ________

Kicking Leg: Left - Right - Alt. **Kicking Height: Low - Mid - High**

Notes:____

____

____

Slow Reps: 10 - 20 - 30 - 40 - 50 - 60 - 70 - 80 - 90 - 100 - 110 - 120 - 130
140 - 150 - 160 - 170 - 180 - 190 - 200 - 210 - 220 - 230 - 240 - 250 - 260
270 - 280 - 290 - 300 - 310 - 320 - 330 - 340 - 350 - 360 - 370 - 380 - 390
400 - 410 - 420 - 430 - 440 - 450 - 460 - 470 - 480 - 490 - 500 - 510 - 520
530 - 540 - 550 - 560 - 570 - 580 - 590 - 600 - 610 - 620 - 630 - 640 - 650
660 - 670 - 680 - 690 - 700 - 710 - 720 - 730 - 740 - 750 - 760 - 770 - 780
790 - 800 - 810 - 820 - 830 - 840 - 850 - 860 - 870 - 880 - 890 - 900 - 910
920 - 930 - 940 - 950 - 960 - 970 - 980 - 990 - 1000

Reg Reps: 10 - 20 - 30 - 40 - 50 - 60 - 70 - 80 - 90 - 100 - 110 - 120 - 130
140 - 150 - 160 - 170 - 180 - 190 - 200 - 210 - 220 - 230 - 240 - 250 - 260
270 - 280 - 290 - 300 - 310 - 320 - 330 - 340 - 350 - 360 - 370 - 380 - 390
400 - 410 - 420 - 430 - 440 - 450 - 460 - 470 - 480 - 490 - 500 - 510 - 520
530 - 540 - 550 - 560 - 570 - 580 - 590 - 600 - 610 - 620 - 630 - 640 - 650
660 - 670 - 680 - 690 - 700 - 710 - 720 - 730 - 740 - 750 - 760 - 770 - 780
790 - 800 - 810 - 820 - 830 - 840 - 850 - 860 - 870 - 880 - 890 - 900 - 910
920 - 930 - 940 - 950 - 960 - 970 - 980 - 990 - 1000

Notes: ____

____

____

____

Reg Reps: 10 - 20 - 30 - 40 - 50 - 60 - 70 - 80 - 90 - 100 - 110 - 120 - 130
140 - 150 - 160 - 170 - 180 - 190 - 200 - 210 - 220 - 230 - 240 - 250 - 260
270 - 280 - 290 - 300 - 310 - 320 - 330 - 340 - 350 - 360 - 370 - 380 - 390
400 - 410 - 420 - 430 - 440 - 450 - 460 - 470 - 480 - 490 - 500 - 510 - 520
530 - 540 - 550 - 560 - 570 - 580 - 590 - 600 - 610 - 620 - 630 - 640 - 650
660 - 670 - 680 - 690 - 700 - 710 - 720 - 730 - 740 - 750 - 760 - 770 - 780
790 - 800 - 810 - 820 - 830 - 840 - 850 - 860 - 870 - 880 - 890 - 900 - 910
920 - 930 - 940 - 950 - 960 - 970 - 980 - 990 - 1000

Notes: ____

____

____

____

Reg Reps: 10 - 20 - 30 - 40 - 50 - 60 - 70 - 80 - 90 - 100 - 110 - 120 - 130
140 - 150 - 160 - 170 - 180 - 190 - 200 - 210 - 220 - 230 - 240 - 250 - 260
270 - 280 - 290 - 300 - 310 - 320 - 330 - 340 - 350 - 360 - 370 - 380 - 390
400 - 410 - 420 - 430 - 440 - 450 - 460 - 470 - 480 - 490 - 500 - 510 - 520
530 - 540 - 550 - 560 - 570 - 580 - 590 - 600 - 610 - 620 - 630 - 640 - 650
660 - 670 - 680 - 690 - 700 - 710 - 720 - 730 - 740 - 750 - 760 - 770 - 780
790 - 800 - 810 - 820 - 830 - 840 - 850 - 860 - 870 - 880 - 890 - 900 - 910
920 - 930 - 940 - 950 - 960 - 970 - 980 - 990 - 1000

Notes: ____

____

____

____

Reg Reps: 10 - 20 - 30 - 40 - 50 - 60 - 70 - 80 - 90 - 100 - 110 - 120 - 130
140 - 150 - 160 - 170 - 180 - 190 - 200 - 210 - 220 - 230 - 240 - 250 - 260
270 - 280 - 290 - 300 - 310 - 320 - 330 - 340 - 350 - 360 - 370 - 380 - 390
400 - 410 - 420 - 430 - 440 - 450 - 460 - 470 - 480 - 490 - 500 - 510 - 520
530 - 540 - 550 - 560 - 570 - 580 - 590 - 600 - 610 - 620 - 630 - 640 - 650
660 - 670 - 680 - 690 - 700 - 710 - 720 - 730 - 740 - 750 - 760 - 770 - 780
790 - 800 - 810 - 820 - 830 - 840 - 850 - 860 - 870 - 880 - 890 - 900 - 910
920 - 930 - 940 - 950 - 960 - 970 - 980 - 990 - 1000

Notes: ____

____

____

____

Reg Reps: 10 - 20 - 30 - 40 - 50 - 60 - 70 - 80 - 90 - 100 - 110 - 120 - 130
140 - 150 - 160 - 170 - 180 - 190 - 200 - 210 - 220 - 230 - 240 - 250 - 260
270 - 280 - 290 - 300 - 310 - 320 - 330 - 340 - 350 - 360 - 370 - 380 - 390
400 - 410 - 420 - 430 - 440 - 450 - 460 - 470 - 480 - 490 - 500 - 510 - 520
530 - 540 - 550 - 560 - 570 - 580 - 590 - 600 - 610 - 620 - 630 - 640 - 650
660 - 670 - 680 - 690 - 700 - 710 - 720 - 730 - 740 - 750 - 760 - 770 - 780
790 - 800 - 810 - 820 - 830 - 840 - 850 - 860 - 870 - 880 - 890 - 900 - 910
920 - 930 - 940 - 950 - 960 - 970 - 980 - 990 - 1000

Notes: ____

____

____

______________________________ **End Date:** ____________

Kick:________________________________ **Start Date:** _________

Kicking Leg: Left - Right - Alt. **Kicking Height: Low - Mid - High**

Notes:____

Slow Reps: 10 - 20 - 30 - 40 - 50 - 60 - 70 - 80 - 90 - 100 - 110 - 120 - 130
140 - 150 - 160 - 170 - 180 - 190 - 200 - 210 - 220 - 230 - 240 - 250 - 260
270 - 280 - 290 - 300 - 310 - 320 - 330 - 340 - 350 - 360 - 370 - 380 - 390
400 - 410 - 420 - 430 - 440 - 450 - 460 - 470 - 480 - 490 - 500 - 510 - 520
530 - 540 - 550 - 560 - 570 - 580 - 590 - 600 - 610 - 620 - 630 - 640 - 650
660 - 670 - 680 - 690 - 700 - 710 - 720 - 730 - 740 - 750 - 760 - 770 - 780
790 - 800 - 810 - 820 - 830 - 840 - 850 - 860 - 870 - 880 - 890 - 900 - 910
920 - 930 - 940 - 950 - 960 - 970 - 980 - 990 - 1000

Reg Reps: 10 - 20 - 30 - 40 - 50 - 60 - 70 - 80 - 90 - 100 - 110 - 120 - 130
140 - 150 - 160 - 170 - 180 - 190 - 200 - 210 - 220 - 230 - 240 - 250 - 260
270 - 280 - 290 - 300 - 310 - 320 - 330 - 340 - 350 - 360 - 370 - 380 - 390
400 - 410 - 420 - 430 - 440 - 450 - 460 - 470 - 480 - 490 - 500 - 510 - 520
530 - 540 - 550 - 560 - 570 - 580 - 590 - 600 - 610 - 620 - 630 - 640 - 650
660 - 670 - 680 - 690 - 700 - 710 - 720 - 730 - 740 - 750 - 760 - 770 - 780
790 - 800 - 810 - 820 - 830 - 840 - 850 - 860 - 870 - 880 - 890 - 900 - 910
920 - 930 - 940 - 950 - 960 - 970 - 980 - 990 - 1000

Notes: ____

Reg Reps: 10 - 20 - 30 - 40 - 50 - 60 - 70 - 80 - 90 - 100 - 110 - 120 - 130
140 - 150 - 160 - 170 - 180 - 190 - 200 - 210 - 220 - 230 - 240 - 250 - 260
270 - 280 - 290 - 300 - 310 - 320 - 330 - 340 - 350 - 360 - 370 - 380 - 390
400 - 410 - 420 - 430 - 440 - 450 - 460 - 470 - 480 - 490 - 500 - 510 - 520
530 - 540 - 550 - 560 - 570 - 580 - 590 - 600 - 610 - 620 - 630 - 640 - 650
660 - 670 - 680 - 690 - 700 - 710 - 720 - 730 - 740 - 750 - 760 - 770 - 780
790 - 800 - 810 - 820 - 830 - 840 - 850 - 860 - 870 - 880 - 890 - 900 - 910
920 - 930 - 940 - 950 - 960 - 970 - 980 - 990 - 1000

Notes: ____

Reg Reps: 10 - 20 - 30 - 40 - 50 - 60 - 70 - 80 - 90 - 100 - 110 - 120 - 130 140 - 150 - 160 - 170 - 180 - 190 - 200 - 210 - 220 - 230 - 240 - 250 - 260 270 - 280 - 290 - 300 - 310 - 320 - 330 - 340 - 350 - 360 - 370 - 380 - 390 400 - 410 - 420 - 430 - 440 - 450 - 460 - 470 - 480 - 490 - 500 - 510 - 520 530 - 540 - 550 - 560 - 570 - 580 - 590 - 600 - 610 - 620 - 630 - 640 - 650 660 - 670 - 680 - 690 - 700 - 710 - 720 - 730 - 740 - 750 - 760 - 770 - 780 790 - 800 - 810 - 820 - 830 - 840 - 850 - 860 - 870 - 880 - 890 - 900 - 910 920 - 930 - 940 - 950 - 960 - 970 - 980 - 990 - 1000

Notes: ____

____

____

____

Reg Reps: 10 - 20 - 30 - 40 - 50 - 60 - 70 - 80 - 90 - 100 - 110 - 120 - 130 140 - 150 - 160 - 170 - 180 - 190 - 200 - 210 - 220 - 230 - 240 - 250 - 260 270 - 280 - 290 - 300 - 310 - 320 - 330 - 340 - 350 - 360 - 370 - 380 - 390 400 - 410 - 420 - 430 - 440 - 450 - 460 - 470 - 480 - 490 - 500 - 510 - 520 530 - 540 - 550 - 560 - 570 - 580 - 590 - 600 - 610 - 620 - 630 - 640 - 650 660 - 670 - 680 - 690 - 700 - 710 - 720 - 730 - 740 - 750 - 760 - 770 - 780 790 - 800 - 810 - 820 - 830 - 840 - 850 - 860 - 870 - 880 - 890 - 900 - 910 920 - 930 - 940 - 950 - 960 - 970 - 980 - 990 - 1000

Notes: ____

____

____

____

Reg Reps: 10 - 20 - 30 - 40 - 50 - 60 - 70 - 80 - 90 - 100 - 110 - 120 - 130 140 - 150 - 160 - 170 - 180 - 190 - 200 - 210 - 220 - 230 - 240 - 250 - 260 270 - 280 - 290 - 300 - 310 - 320 - 330 - 340 - 350 - 360 - 370 - 380 - 390 400 - 410 - 420 - 430 - 440 - 450 - 460 - 470 - 480 - 490 - 500 - 510 - 520 530 - 540 - 550 - 560 - 570 - 580 - 590 - 600 - 610 - 620 - 630 - 640 - 650 660 - 670 - 680 - 690 - 700 - 710 - 720 - 730 - 740 - 750 - 760 - 770 - 780 790 - 800 - 810 - 820 - 830 - 840 - 850 - 860 - 870 - 880 - 890 - 900 - 910 920 - 930 - 940 - 950 - 960 - 970 - 980 - 990 - 1000

Notes: ____

____

____

______________________________ **End Date:** ____________

Kick:______________________________ **Start Date:** __________

Kicking Leg: Left - Right - Alt. **Kicking Height: Low - Mid - High**

Notes:____

____

____

Slow Reps: 10 - 20 - 30 - 40 - 50 - 60 - 70 - 80 - 90 - 100 - 110 - 120 - 130
140 - 150 - 160 - 170 - 180 - 190 - 200 - 210 - 220 - 230 - 240 - 250 - 260
270 - 280 - 290 - 300 - 310 - 320 - 330 - 340 - 350 - 360 - 370 - 380 - 390
400 - 410 - 420 - 430 - 440 - 450 - 460 - 470 - 480 - 490 - 500 - 510 - 520
530 - 540 - 550 - 560 - 570 - 580 - 590 - 600 - 610 - 620 - 630 - 640 - 650
660 - 670 - 680 - 690 - 700 - 710 - 720 - 730 - 740 - 750 - 760 - 770 - 780
790 - 800 - 810 - 820 - 830 - 840 - 850 - 860 - 870 - 880 - 890 - 900 - 910
920 - 930 - 940 - 950 - 960 - 970 - 980 - 990 - 1000

Reg Reps: 10 - 20 - 30 - 40 - 50 - 60 - 70 - 80 - 90 - 100 - 110 - 120 - 130
140 - 150 - 160 - 170 - 180 - 190 - 200 - 210 - 220 - 230 - 240 - 250 - 260
270 - 280 - 290 - 300 - 310 - 320 - 330 - 340 - 350 - 360 - 370 - 380 - 390
400 - 410 - 420 - 430 - 440 - 450 - 460 - 470 - 480 - 490 - 500 - 510 - 520
530 - 540 - 550 - 560 - 570 - 580 - 590 - 600 - 610 - 620 - 630 - 640 - 650
660 - 670 - 680 - 690 - 700 - 710 - 720 - 730 - 740 - 750 - 760 - 770 - 780
790 - 800 - 810 - 820 - 830 - 840 - 850 - 860 - 870 - 880 - 890 - 900 - 910
920 - 930 - 940 - 950 - 960 - 970 - 980 - 990 - 1000

Notes: ____

____

____

____

Reg Reps: 10 - 20 - 30 - 40 - 50 - 60 - 70 - 80 - 90 - 100 - 110 - 120 - 130
140 - 150 - 160 - 170 - 180 - 190 - 200 - 210 - 220 - 230 - 240 - 250 - 260
270 - 280 - 290 - 300 - 310 - 320 - 330 - 340 - 350 - 360 - 370 - 380 - 390
400 - 410 - 420 - 430 - 440 - 450 - 460 - 470 - 480 - 490 - 500 - 510 - 520
530 - 540 - 550 - 560 - 570 - 580 - 590 - 600 - 610 - 620 - 630 - 640 - 650
660 - 670 - 680 - 690 - 700 - 710 - 720 - 730 - 740 - 750 - 760 - 770 - 780
790 - 800 - 810 - 820 - 830 - 840 - 850 - 860 - 870 - 880 - 890 - 900 - 910
920 - 930 - 940 - 950 - 960 - 970 - 980 - 990 - 1000

Notes: ____

____

____

____

Reg Reps: 10 - 20 - 30 - 40 - 50 - 60 - 70 - 80 - 90 - 100 - 110 - 120 - 130
140 - 150 - 160 - 170 - 180 - 190 - 200 - 210 - 220 - 230 - 240 - 250 - 260
270 - 280 - 290 - 300 - 310 - 320 - 330 - 340 - 350 - 360 - 370 - 380 - 390
400 - 410 - 420 - 430 - 440 - 450 - 460 - 470 - 480 - 490 - 500 - 510 - 520
530 - 540 - 550 - 560 - 570 - 580 - 590 - 600 - 610 - 620 - 630 - 640 - 650
660 - 670 - 680 - 690 - 700 - 710 - 720 - 730 - 740 - 750 - 760 - 770 - 780
790 - 800 - 810 - 820 - 830 - 840 - 850 - 860 - 870 - 880 - 890 - 900 - 910
920 - 930 - 940 - 950 - 960 - 970 - 980 - 990 - 1000

Notes: ____

____

____

____

Reg Reps: 10 - 20 - 30 - 40 - 50 - 60 - 70 - 80 - 90 - 100 - 110 - 120 - 130
140 - 150 - 160 - 170 - 180 - 190 - 200 - 210 - 220 - 230 - 240 - 250 - 260
270 - 280 - 290 - 300 - 310 - 320 - 330 - 340 - 350 - 360 - 370 - 380 - 390
400 - 410 - 420 - 430 - 440 - 450 - 460 - 470 - 480 - 490 - 500 - 510 - 520
530 - 540 - 550 - 560 - 570 - 580 - 590 - 600 - 610 - 620 - 630 - 640 - 650
660 - 670 - 680 - 690 - 700 - 710 - 720 - 730 - 740 - 750 - 760 - 770 - 780
790 - 800 - 810 - 820 - 830 - 840 - 850 - 860 - 870 - 880 - 890 - 900 - 910
920 - 930 - 940 - 950 - 960 - 970 - 980 - 990 - 1000

Notes: ____

____

____

____

Reg Reps: 10 - 20 - 30 - 40 - 50 - 60 - 70 - 80 - 90 - 100 - 110 - 120 - 130
140 - 150 - 160 - 170 - 180 - 190 - 200 - 210 - 220 - 230 - 240 - 250 - 260
270 - 280 - 290 - 300 - 310 - 320 - 330 - 340 - 350 - 360 - 370 - 380 - 390
400 - 410 - 420 - 430 - 440 - 450 - 460 - 470 - 480 - 490 - 500 - 510 - 520
530 - 540 - 550 - 560 - 570 - 580 - 590 - 600 - 610 - 620 - 630 - 640 - 650
660 - 670 - 680 - 690 - 700 - 710 - 720 - 730 - 740 - 750 - 760 - 770 - 780
790 - 800 - 810 - 820 - 830 - 840 - 850 - 860 - 870 - 880 - 890 - 900 - 910
920 - 930 - 940 - 950 - 960 - 970 - 980 - 990 - 1000

Notes: ____

____

____

______________________________ **End Date:** ____________

Kick: ______________________________ **Start Date:** __________

Kicking Leg: Left - Right - Alt. **Kicking Height: Low - Mid - High**

Notes: ____

____

____

Slow Reps: 10 - 20 - 30 - 40 - 50 - 60 - 70 - 80 - 90 - 100 - 110 - 120 - 130
140 - 150 - 160 - 170 - 180 - 190 - 200 - 210 - 220 - 230 - 240 - 250 - 260
270 - 280 - 290 - 300 - 310 - 320 - 330 - 340 - 350 - 360 - 370 - 380 - 390
400 - 410 - 420 - 430 - 440 - 450 - 460 - 470 - 480 - 490 - 500 - 510 - 520
530 - 540 - 550 - 560 - 570 - 580 - 590 - 600 - 610 - 620 - 630 - 640 - 650
660 - 670 - 680 - 690 - 700 - 710 - 720 - 730 - 740 - 750 - 760 - 770 - 780
790 - 800 - 810 - 820 - 830 - 840 - 850 - 860 - 870 - 880 - 890 - 900 - 910
920 - 930 - 940 - 950 - 960 - 970 - 980 - 990 - 1000

Reg Reps: 10 - 20 - 30 - 40 - 50 - 60 - 70 - 80 - 90 - 100 - 110 - 120 - 130
140 - 150 - 160 - 170 - 180 - 190 - 200 - 210 - 220 - 230 - 240 - 250 - 260
270 - 280 - 290 - 300 - 310 - 320 - 330 - 340 - 350 - 360 - 370 - 380 - 390
400 - 410 - 420 - 430 - 440 - 450 - 460 - 470 - 480 - 490 - 500 - 510 - 520
530 - 540 - 550 - 560 - 570 - 580 - 590 - 600 - 610 - 620 - 630 - 640 - 650
660 - 670 - 680 - 690 - 700 - 710 - 720 - 730 - 740 - 750 - 760 - 770 - 780
790 - 800 - 810 - 820 - 830 - 840 - 850 - 860 - 870 - 880 - 890 - 900 - 910
920 - 930 - 940 - 950 - 960 - 970 - 980 - 990 - 1000

Notes: ____

____

____

____

Reg Reps: 10 - 20 - 30 - 40 - 50 - 60 - 70 - 80 - 90 - 100 - 110 - 120 - 130
140 - 150 - 160 - 170 - 180 - 190 - 200 - 210 - 220 - 230 - 240 - 250 - 260
270 - 280 - 290 - 300 - 310 - 320 - 330 - 340 - 350 - 360 - 370 - 380 - 390
400 - 410 - 420 - 430 - 440 - 450 - 460 - 470 - 480 - 490 - 500 - 510 - 520
530 - 540 - 550 - 560 - 570 - 580 - 590 - 600 - 610 - 620 - 630 - 640 - 650
660 - 670 - 680 - 690 - 700 - 710 - 720 - 730 - 740 - 750 - 760 - 770 - 780
790 - 800 - 810 - 820 - 830 - 840 - 850 - 860 - 870 - 880 - 890 - 900 - 910
920 - 930 - 940 - 950 - 960 - 970 - 980 - 990 - 1000

Notes: ____

____

____

____

Reg Reps: 10 - 20 - 30 - 40 - 50 - 60 - 70 - 80 - 90 - 100 - 110 - 120 - 130
140 - 150 - 160 - 170 - 180 - 190 - 200 - 210 - 220 - 230 - 240 - 250 - 260
270 - 280 - 290 - 300 - 310 - 320 - 330 - 340 - 350 - 360 - 370 - 380 - 390
400 - 410 - 420 - 430 - 440 - 450 - 460 - 470 - 480 - 490 - 500 - 510 - 520
530 - 540 - 550 - 560 - 570 - 580 - 590 - 600 - 610 - 620 - 630 - 640 - 650
660 - 670 - 680 - 690 - 700 - 710 - 720 - 730 - 740 - 750 - 760 - 770 - 780
790 - 800 - 810 - 820 - 830 - 840 - 850 - 860 - 870 - 880 - 890 - 900 - 910
920 - 930 - 940 - 950 - 960 - 970 - 980 - 990 - 1000

Notes: ____

____

____

____

Reg Reps: 10 - 20 - 30 - 40 - 50 - 60 - 70 - 80 - 90 - 100 - 110 - 120 - 130
140 - 150 - 160 - 170 - 180 - 190 - 200 - 210 - 220 - 230 - 240 - 250 - 260
270 - 280 - 290 - 300 - 310 - 320 - 330 - 340 - 350 - 360 - 370 - 380 - 390
400 - 410 - 420 - 430 - 440 - 450 - 460 - 470 - 480 - 490 - 500 - 510 - 520
530 - 540 - 550 - 560 - 570 - 580 - 590 - 600 - 610 - 620 - 630 - 640 - 650
660 - 670 - 680 - 690 - 700 - 710 - 720 - 730 - 740 - 750 - 760 - 770 - 780
790 - 800 - 810 - 820 - 830 - 840 - 850 - 860 - 870 - 880 - 890 - 900 - 910
920 - 930 - 940 - 950 - 960 - 970 - 980 - 990 - 1000

Notes: ____

____

____

____

Reg Reps: 10 - 20 - 30 - 40 - 50 - 60 - 70 - 80 - 90 - 100 - 110 - 120 - 130
140 - 150 - 160 - 170 - 180 - 190 - 200 - 210 - 220 - 230 - 240 - 250 - 260
270 - 280 - 290 - 300 - 310 - 320 - 330 - 340 - 350 - 360 - 370 - 380 - 390
400 - 410 - 420 - 430 - 440 - 450 - 460 - 470 - 480 - 490 - 500 - 510 - 520
530 - 540 - 550 - 560 - 570 - 580 - 590 - 600 - 610 - 620 - 630 - 640 - 650
660 - 670 - 680 - 690 - 700 - 710 - 720 - 730 - 740 - 750 - 760 - 770 - 780
790 - 800 - 810 - 820 - 830 - 840 - 850 - 860 - 870 - 880 - 890 - 900 - 910
920 - 930 - 940 - 950 - 960 - 970 - 980 - 990 - 1000

Notes: ____

____

____

________________________________ **End Date:** ____________

Kick:____**Start Date:** __________

Kicking Leg: Left - Right - Alt. **Kicking Height: Low - Mid - High**

Notes:____

____

____

Slow Reps: 10 - 20 - 30 - 40 - 50 - 60 - 70 - 80 - 90 - 100 - 110 - 120 - 130
140 - 150 - 160 - 170 - 180 - 190 - 200 - 210 - 220 - 230 - 240 - 250 - 260
270 - 280 - 290 - 300 - 310 - 320 - 330 - 340 - 350 - 360 - 370 - 380 - 390
400 - 410 - 420 - 430 - 440 - 450 - 460 - 470 - 480 - 490 - 500 - 510 - 520
530 - 540 - 550 - 560 - 570 - 580 - 590 - 600 - 610 - 620 - 630 - 640 - 650
660 - 670 - 680 - 690 - 700 - 710 - 720 - 730 - 740 - 750 - 760 - 770 - 780
790 - 800 - 810 - 820 - 830 - 840 - 850 - 860 - 870 - 880 - 890 - 900 - 910
920 - 930 - 940 - 950 - 960 - 970 - 980 - 990 - 1000

Reg Reps: 10 - 20 - 30 - 40 - 50 - 60 - 70 - 80 - 90 - 100 - 110 - 120 - 130
140 - 150 - 160 - 170 - 180 - 190 - 200 - 210 - 220 - 230 - 240 - 250 - 260
270 - 280 - 290 - 300 - 310 - 320 - 330 - 340 - 350 - 360 - 370 - 380 - 390
400 - 410 - 420 - 430 - 440 - 450 - 460 - 470 - 480 - 490 - 500 - 510 - 520
530 - 540 - 550 - 560 - 570 - 580 - 590 - 600 - 610 - 620 - 630 - 640 - 650
660 - 670 - 680 - 690 - 700 - 710 - 720 - 730 - 740 - 750 - 760 - 770 - 780
790 - 800 - 810 - 820 - 830 - 840 - 850 - 860 - 870 - 880 - 890 - 900 - 910
920 - 930 - 940 - 950 - 960 - 970 - 980 - 990 - 1000

Notes: ____

____

____

____

Reg Reps: 10 - 20 - 30 - 40 - 50 - 60 - 70 - 80 - 90 - 100 - 110 - 120 - 130
140 - 150 - 160 - 170 - 180 - 190 - 200 - 210 - 220 - 230 - 240 - 250 - 260
270 - 280 - 290 - 300 - 310 - 320 - 330 - 340 - 350 - 360 - 370 - 380 - 390
400 - 410 - 420 - 430 - 440 - 450 - 460 - 470 - 480 - 490 - 500 - 510 - 520
530 - 540 - 550 - 560 - 570 - 580 - 590 - 600 - 610 - 620 - 630 - 640 - 650
660 - 670 - 680 - 690 - 700 - 710 - 720 - 730 - 740 - 750 - 760 - 770 - 780
790 - 800 - 810 - 820 - 830 - 840 - 850 - 860 - 870 - 880 - 890 - 900 - 910
920 - 930 - 940 - 950 - 960 - 970 - 980 - 990 - 1000

Notes: ____

____

____

____

Reg Reps: 10 - 20 - 30 - 40 - 50 - 60 - 70 - 80 - 90 - 100 - 110 - 120 - 130 - 140 - 150 - 160 - 170 - 180 - 190 - 200 - 210 - 220 - 230 - 240 - 250 - 260 - 270 - 280 - 290 - 300 - 310 - 320 - 330 - 340 - 350 - 360 - 370 - 380 - 390 - 400 - 410 - 420 - 430 - 440 - 450 - 460 - 470 - 480 - 490 - 500 - 510 - 520 - 530 - 540 - 550 - 560 - 570 - 580 - 590 - 600 - 610 - 620 - 630 - 640 - 650 - 660 - 670 - 680 - 690 - 700 - 710 - 720 - 730 - 740 - 750 - 760 - 770 - 780 - 790 - 800 - 810 - 820 - 830 - 840 - 850 - 860 - 870 - 880 - 890 - 900 - 910 - 920 - 930 - 940 - 950 - 960 - 970 - 980 - 990 - 1000

Notes: ____

____

____

____

Reg Reps: 10 - 20 - 30 - 40 - 50 - 60 - 70 - 80 - 90 - 100 - 110 - 120 - 130 - 140 - 150 - 160 - 170 - 180 - 190 - 200 - 210 - 220 - 230 - 240 - 250 - 260 - 270 - 280 - 290 - 300 - 310 - 320 - 330 - 340 - 350 - 360 - 370 - 380 - 390 - 400 - 410 - 420 - 430 - 440 - 450 - 460 - 470 - 480 - 490 - 500 - 510 - 520 - 530 - 540 - 550 - 560 - 570 - 580 - 590 - 600 - 610 - 620 - 630 - 640 - 650 - 660 - 670 - 680 - 690 - 700 - 710 - 720 - 730 - 740 - 750 - 760 - 770 - 780 - 790 - 800 - 810 - 820 - 830 - 840 - 850 - 860 - 870 - 880 - 890 - 900 - 910 - 920 - 930 - 940 - 950 - 960 - 970 - 980 - 990 - 1000

Notes: ____

____

____

____

Reg Reps: 10 - 20 - 30 - 40 - 50 - 60 - 70 - 80 - 90 - 100 - 110 - 120 - 130 - 140 - 150 - 160 - 170 - 180 - 190 - 200 - 210 - 220 - 230 - 240 - 250 - 260 - 270 - 280 - 290 - 300 - 310 - 320 - 330 - 340 - 350 - 360 - 370 - 380 - 390 - 400 - 410 - 420 - 430 - 440 - 450 - 460 - 470 - 480 - 490 - 500 - 510 - 520 - 530 - 540 - 550 - 560 - 570 - 580 - 590 - 600 - 610 - 620 - 630 - 640 - 650 - 660 - 670 - 680 - 690 - 700 - 710 - 720 - 730 - 740 - 750 - 760 - 770 - 780 - 790 - 800 - 810 - 820 - 830 - 840 - 850 - 860 - 870 - 880 - 890 - 900 - 910 - 920 - 930 - 940 - 950 - 960 - 970 - 980 - 990 - 1000

Notes: ____

____

____

______________________________ **End Date:** ____________

Kick:________________________________ **Start Date:** __________
Kicking Leg: Left - Right - Alt. **Kicking Height: Low - Mid - High**
Notes:____

Slow Reps: 10 - 20 - 30 - 40 - 50 - 60 - 70 - 80 - 90 - 100 - 110 - 120 - 130
140 - 150 - 160 - 170 - 180 - 190 - 200 - 210 - 220 - 230 - 240 - 250 - 260
270 - 280 - 290 - 300 - 310 - 320 - 330 - 340 - 350 - 360 - 370 - 380 - 390
400 - 410 - 420 - 430 - 440 - 450 - 460 - 470 - 480 - 490 - 500 - 510 - 520
530 - 540 - 550 - 560 - 570 - 580 - 590 - 600 - 610 - 620 - 630 - 640 - 650
660 - 670 - 680 - 690 - 700 - 710 - 720 - 730 - 740 - 750 - 760 - 770 - 780
790 - 800 - 810 - 820 - 830 - 840 - 850 - 860 - 870 - 880 - 890 - 900 - 910
920 - 930 - 940 - 950 - 960 - 970 - 980 - 990 - 1000

Reg Reps: 10 - 20 - 30 - 40 - 50 - 60 - 70 - 80 - 90 - 100 - 110 - 120 - 130
140 - 150 - 160 - 170 - 180 - 190 - 200 - 210 - 220 - 230 - 240 - 250 - 260
270 - 280 - 290 - 300 - 310 - 320 - 330 - 340 - 350 - 360 - 370 - 380 - 390
400 - 410 - 420 - 430 - 440 - 450 - 460 - 470 - 480 - 490 - 500 - 510 - 520
530 - 540 - 550 - 560 - 570 - 580 - 590 - 600 - 610 - 620 - 630 - 640 - 650
660 - 670 - 680 - 690 - 700 - 710 - 720 - 730 - 740 - 750 - 760 - 770 - 780
790 - 800 - 810 - 820 - 830 - 840 - 850 - 860 - 870 - 880 - 890 - 900 - 910
920 - 930 - 940 - 950 - 960 - 970 - 980 - 990 - 1000
Notes: ____

Reg Reps: 10 - 20 - 30 - 40 - 50 - 60 - 70 - 80 - 90 - 100 - 110 - 120 - 130
140 - 150 - 160 - 170 - 180 - 190 - 200 - 210 - 220 - 230 - 240 - 250 - 260
270 - 280 - 290 - 300 - 310 - 320 - 330 - 340 - 350 - 360 - 370 - 380 - 390
400 - 410 - 420 - 430 - 440 - 450 - 460 - 470 - 480 - 490 - 500 - 510 - 520
530 - 540 - 550 - 560 - 570 - 580 - 590 - 600 - 610 - 620 - 630 - 640 - 650
660 - 670 - 680 - 690 - 700 - 710 - 720 - 730 - 740 - 750 - 760 - 770 - 780
790 - 800 - 810 - 820 - 830 - 840 - 850 - 860 - 870 - 880 - 890 - 900 - 910
920 - 930 - 940 - 950 - 960 - 970 - 980 - 990 - 1000
Notes: ____

Reg Reps: 10 - 20 - 30 - 40 - 50 - 60 - 70 - 80 - 90 - 100 - 110 - 120 - 130
140 - 150 - 160 - 170 - 180 - 190 - 200 - 210 - 220 - 230 - 240 - 250 - 260
270 - 280 - 290 - 300 - 310 - 320 - 330 - 340 - 350 - 360 - 370 - 380 - 390
400 - 410 - 420 - 430 - 440 - 450 - 460 - 470 - 480 - 490 - 500 - 510 - 520
530 - 540 - 550 - 560 - 570 - 580 - 590 - 600 - 610 - 620 - 630 - 640 - 650
660 - 670 - 680 - 690 - 700 - 710 - 720 - 730 - 740 - 750 - 760 - 770 - 780
790 - 800 - 810 - 820 - 830 - 840 - 850 - 860 - 870 - 880 - 890 - 900 - 910
920 - 930 - 940 - 950 - 960 - 970 - 980 - 990 - 1000

Notes: ____

____

____

____

Reg Reps: 10 - 20 - 30 - 40 - 50 - 60 - 70 - 80 - 90 - 100 - 110 - 120 - 130
140 - 150 - 160 - 170 - 180 - 190 - 200 - 210 - 220 - 230 - 240 - 250 - 260
270 - 280 - 290 - 300 - 310 - 320 - 330 - 340 - 350 - 360 - 370 - 380 - 390
400 - 410 - 420 - 430 - 440 - 450 - 460 - 470 - 480 - 490 - 500 - 510 - 520
530 - 540 - 550 - 560 - 570 - 580 - 590 - 600 - 610 - 620 - 630 - 640 - 650
660 - 670 - 680 - 690 - 700 - 710 - 720 - 730 - 740 - 750 - 760 - 770 - 780
790 - 800 - 810 - 820 - 830 - 840 - 850 - 860 - 870 - 880 - 890 - 900 - 910
920 - 930 - 940 - 950 - 960 - 970 - 980 - 990 - 1000

Notes: ____

____

____

____

Reg Reps: 10 - 20 - 30 - 40 - 50 - 60 - 70 - 80 - 90 - 100 - 110 - 120 - 130
140 - 150 - 160 - 170 - 180 - 190 - 200 - 210 - 220 - 230 - 240 - 250 - 260
270 - 280 - 290 - 300 - 310 - 320 - 330 - 340 - 350 - 360 - 370 - 380 - 390
400 - 410 - 420 - 430 - 440 - 450 - 460 - 470 - 480 - 490 - 500 - 510 - 520
530 - 540 - 550 - 560 - 570 - 580 - 590 - 600 - 610 - 620 - 630 - 640 - 650
660 - 670 - 680 - 690 - 700 - 710 - 720 - 730 - 740 - 750 - 760 - 770 - 780
790 - 800 - 810 - 820 - 830 - 840 - 850 - 860 - 870 - 880 - 890 - 900 - 910
920 - 930 - 940 - 950 - 960 - 970 - 980 - 990 - 1000

Notes: ____

____

____

________________________________ **End Date:** ____________

Kick: ______________________________ **Start Date:** ________

Kicking Leg: Left - Right - Alt. **Kicking Height: Low - Mid - High**

Notes: ____

____

____

Slow Reps: 10 - 20 - 30 - 40 - 50 - 60 - 70 - 80 - 90 - 100 - 110 - 120 - 130
140 - 150 - 160 - 170 - 180 - 190 - 200 - 210 - 220 - 230 - 240 - 250 - 260
270 - 280 - 290 - 300 - 310 - 320 - 330 - 340 - 350 - 360 - 370 - 380 - 390
400 - 410 - 420 - 430 - 440 - 450 - 460 - 470 - 480 - 490 - 500 - 510 - 520
530 - 540 - 550 - 560 - 570 - 580 - 590 - 600 - 610 - 620 - 630 - 640 - 650
660 - 670 - 680 - 690 - 700 - 710 - 720 - 730 - 740 - 750 - 760 - 770 - 780
790 - 800 - 810 - 820 - 830 - 840 - 850 - 860 - 870 - 880 - 890 - 900 - 910
920 - 930 - 940 - 950 - 960 - 970 - 980 - 990 - 1000

Reg Reps: 10 - 20 - 30 - 40 - 50 - 60 - 70 - 80 - 90 - 100 - 110 - 120 - 130
140 - 150 - 160 - 170 - 180 - 190 - 200 - 210 - 220 - 230 - 240 - 250 - 260
270 - 280 - 290 - 300 - 310 - 320 - 330 - 340 - 350 - 360 - 370 - 380 - 390
400 - 410 - 420 - 430 - 440 - 450 - 460 - 470 - 480 - 490 - 500 - 510 - 520
530 - 540 - 550 - 560 - 570 - 580 - 590 - 600 - 610 - 620 - 630 - 640 - 650
660 - 670 - 680 - 690 - 700 - 710 - 720 - 730 - 740 - 750 - 760 - 770 - 780
790 - 800 - 810 - 820 - 830 - 840 - 850 - 860 - 870 - 880 - 890 - 900 - 910
920 - 930 - 940 - 950 - 960 - 970 - 980 - 990 - 1000

Notes: ____

____

____

____

Reg Reps: 10 - 20 - 30 - 40 - 50 - 60 - 70 - 80 - 90 - 100 - 110 - 120 - 130
140 - 150 - 160 - 170 - 180 - 190 - 200 - 210 - 220 - 230 - 240 - 250 - 260
270 - 280 - 290 - 300 - 310 - 320 - 330 - 340 - 350 - 360 - 370 - 380 - 390
400 - 410 - 420 - 430 - 440 - 450 - 460 - 470 - 480 - 490 - 500 - 510 - 520
530 - 540 - 550 - 560 - 570 - 580 - 590 - 600 - 610 - 620 - 630 - 640 - 650
660 - 670 - 680 - 690 - 700 - 710 - 720 - 730 - 740 - 750 - 760 - 770 - 780
790 - 800 - 810 - 820 - 830 - 840 - 850 - 860 - 870 - 880 - 890 - 900 - 910
920 - 930 - 940 - 950 - 960 - 970 - 980 - 990 - 1000

Notes: ____

____

____

____

Reg Reps: 10 - 20 - 30 - 40 - 50 - 60 - 70 - 80 - 90 - 100 - 110 - 120 - 130 - 140 - 150 - 160 - 170 - 180 - 190 - 200 - 210 - 220 - 230 - 240 - 250 - 260 - 270 - 280 - 290 - 300 - 310 - 320 - 330 - 340 - 350 - 360 - 370 - 380 - 390 - 400 - 410 - 420 - 430 - 440 - 450 - 460 - 470 - 480 - 490 - 500 - 510 - 520 - 530 - 540 - 550 - 560 - 570 - 580 - 590 - 600 - 610 - 620 - 630 - 640 - 650 - 660 - 670 - 680 - 690 - 700 - 710 - 720 - 730 - 740 - 750 - 760 - 770 - 780 - 790 - 800 - 810 - 820 - 830 - 840 - 850 - 860 - 870 - 880 - 890 - 900 - 910 - 920 - 930 - 940 - 950 - 960 - 970 - 980 - 990 - 1000

Notes: ____

____

____

____

Reg Reps: 10 - 20 - 30 - 40 - 50 - 60 - 70 - 80 - 90 - 100 - 110 - 120 - 130 - 140 - 150 - 160 - 170 - 180 - 190 - 200 - 210 - 220 - 230 - 240 - 250 - 260 - 270 - 280 - 290 - 300 - 310 - 320 - 330 - 340 - 350 - 360 - 370 - 380 - 390 - 400 - 410 - 420 - 430 - 440 - 450 - 460 - 470 - 480 - 490 - 500 - 510 - 520 - 530 - 540 - 550 - 560 - 570 - 580 - 590 - 600 - 610 - 620 - 630 - 640 - 650 - 660 - 670 - 680 - 690 - 700 - 710 - 720 - 730 - 740 - 750 - 760 - 770 - 780 - 790 - 800 - 810 - 820 - 830 - 840 - 850 - 860 - 870 - 880 - 890 - 900 - 910 - 920 - 930 - 940 - 950 - 960 - 970 - 980 - 990 - 1000

Notes: ____

____

____

____

Reg Reps: 10 - 20 - 30 - 40 - 50 - 60 - 70 - 80 - 90 - 100 - 110 - 120 - 130 - 140 - 150 - 160 - 170 - 180 - 190 - 200 - 210 - 220 - 230 - 240 - 250 - 260 - 270 - 280 - 290 - 300 - 310 - 320 - 330 - 340 - 350 - 360 - 370 - 380 - 390 - 400 - 410 - 420 - 430 - 440 - 450 - 460 - 470 - 480 - 490 - 500 - 510 - 520 - 530 - 540 - 550 - 560 - 570 - 580 - 590 - 600 - 610 - 620 - 630 - 640 - 650 - 660 - 670 - 680 - 690 - 700 - 710 - 720 - 730 - 740 - 750 - 760 - 770 - 780 - 790 - 800 - 810 - 820 - 830 - 840 - 850 - 860 - 870 - 880 - 890 - 900 - 910 - 920 - 930 - 940 - 950 - 960 - 970 - 980 - 990 - 1000

Notes: ____

____

____

______________________________ **End Date:** ____________

Kick:__________________________________ **Start Date:** _________

Kicking Leg: Left - Right - Alt. **Kicking Height: Low - Mid - High**

Notes:____

____

____

Slow Reps: 10 - 20 - 30 - 40 - 50 - 60 - 70 - 80 - 90 - 100 - 110 - 120 - 130
140 - 150 - 160 - 170 - 180 - 190 - 200 - 210 - 220 - 230 - 240 - 250 - 260
270 - 280 - 290 - 300 - 310 - 320 - 330 - 340 - 350 - 360 - 370 - 380 - 390
400 - 410 - 420 - 430 - 440 - 450 - 460 - 470 - 480 - 490 - 500 - 510 - 520
530 - 540 - 550 - 560 - 570 - 580 - 590 - 600 - 610 - 620 - 630 - 640 - 650
660 - 670 - 680 - 690 - 700 - 710 - 720 - 730 - 740 - 750 - 760 - 770 - 780
790 - 800 - 810 - 820 - 830 - 840 - 850 - 860 - 870 - 880 - 890 - 900 - 910
920 - 930 - 940 - 950 - 960 - 970 - 980 - 990 - 1000

Reg Reps: 10 - 20 - 30 - 40 - 50 - 60 - 70 - 80 - 90 - 100 - 110 - 120 - 130
140 - 150 - 160 - 170 - 180 - 190 - 200 - 210 - 220 - 230 - 240 - 250 - 260
270 - 280 - 290 - 300 - 310 - 320 - 330 - 340 - 350 - 360 - 370 - 380 - 390
400 - 410 - 420 - 430 - 440 - 450 - 460 - 470 - 480 - 490 - 500 - 510 - 520
530 - 540 - 550 - 560 - 570 - 580 - 590 - 600 - 610 - 620 - 630 - 640 - 650
660 - 670 - 680 - 690 - 700 - 710 - 720 - 730 - 740 - 750 - 760 - 770 - 780
790 - 800 - 810 - 820 - 830 - 840 - 850 - 860 - 870 - 880 - 890 - 900 - 910
920 - 930 - 940 - 950 - 960 - 970 - 980 - 990 - 1000

Notes: ____

____

____

____

Reg Reps: 10 - 20 - 30 - 40 - 50 - 60 - 70 - 80 - 90 - 100 - 110 - 120 - 130
140 - 150 - 160 - 170 - 180 - 190 - 200 - 210 - 220 - 230 - 240 - 250 - 260
270 - 280 - 290 - 300 - 310 - 320 - 330 - 340 - 350 - 360 - 370 - 380 - 390
400 - 410 - 420 - 430 - 440 - 450 - 460 - 470 - 480 - 490 - 500 - 510 - 520
530 - 540 - 550 - 560 - 570 - 580 - 590 - 600 - 610 - 620 - 630 - 640 - 650
660 - 670 - 680 - 690 - 700 - 710 - 720 - 730 - 740 - 750 - 760 - 770 - 780
790 - 800 - 810 - 820 - 830 - 840 - 850 - 860 - 870 - 880 - 890 - 900 - 910
920 - 930 - 940 - 950 - 960 - 970 - 980 - 990 - 1000

Notes: ____

____

____

____

Reg Reps: 10 - 20 - 30 - 40 - 50 - 60 - 70 - 80 - 90 - 100 - 110 - 120 - 130 - 140 - 150 - 160 - 170 - 180 - 190 - 200 - 210 - 220 - 230 - 240 - 250 - 260 - 270 - 280 - 290 - 300 - 310 - 320 - 330 - 340 - 350 - 360 - 370 - 380 - 390 - 400 - 410 - 420 - 430 - 440 - 450 - 460 - 470 - 480 - 490 - 500 - 510 - 520 - 530 - 540 - 550 - 560 - 570 - 580 - 590 - 600 - 610 - 620 - 630 - 640 - 650 - 660 - 670 - 680 - 690 - 700 - 710 - 720 - 730 - 740 - 750 - 760 - 770 - 780 - 790 - 800 - 810 - 820 - 830 - 840 - 850 - 860 - 870 - 880 - 890 - 900 - 910 - 920 - 930 - 940 - 950 - 960 - 970 - 980 - 990 - 1000

Notes: ______________________________

____

____

____

Reg Reps: 10 - 20 - 30 - 40 - 50 - 60 - 70 - 80 - 90 - 100 - 110 - 120 - 130 - 140 - 150 - 160 - 170 - 180 - 190 - 200 - 210 - 220 - 230 - 240 - 250 - 260 - 270 - 280 - 290 - 300 - 310 - 320 - 330 - 340 - 350 - 360 - 370 - 380 - 390 - 400 - 410 - 420 - 430 - 440 - 450 - 460 - 470 - 480 - 490 - 500 - 510 - 520 - 530 - 540 - 550 - 560 - 570 - 580 - 590 - 600 - 610 - 620 - 630 - 640 - 650 - 660 - 670 - 680 - 690 - 700 - 710 - 720 - 730 - 740 - 750 - 760 - 770 - 780 - 790 - 800 - 810 - 820 - 830 - 840 - 850 - 860 - 870 - 880 - 890 - 900 - 910 - 920 - 930 - 940 - 950 - 960 - 970 - 980 - 990 - 1000

Notes: ______________________________

____

____

____

Reg Reps: 10 - 20 - 30 - 40 - 50 - 60 - 70 - 80 - 90 - 100 - 110 - 120 - 130 - 140 - 150 - 160 - 170 - 180 - 190 - 200 - 210 - 220 - 230 - 240 - 250 - 260 - 270 - 280 - 290 - 300 - 310 - 320 - 330 - 340 - 350 - 360 - 370 - 380 - 390 - 400 - 410 - 420 - 430 - 440 - 450 - 460 - 470 - 480 - 490 - 500 - 510 - 520 - 530 - 540 - 550 - 560 - 570 - 580 - 590 - 600 - 610 - 620 - 630 - 640 - 650 - 660 - 670 - 680 - 690 - 700 - 710 - 720 - 730 - 740 - 750 - 760 - 770 - 780 - 790 - 800 - 810 - 820 - 830 - 840 - 850 - 860 - 870 - 880 - 890 - 900 - 910 - 920 - 930 - 940 - 950 - 960 - 970 - 980 - 990 - 1000

Notes: ______________________________

____

____

______________________ **End Date:** __________

Kick: ______________________________ **Start Date:** __________

Kicking Leg: Left - Right - Alt. **Kicking Height: Low - Mid - High**

Notes: ____

____

____

Slow Reps: 10 - 20 - 30 - 40 - 50 - 60 - 70 - 80 - 90 - 100 - 110 - 120 - 130
140 - 150 - 160 - 170 - 180 - 190 - 200 - 210 - 220 - 230 - 240 - 250 - 260
270 - 280 - 290 - 300 - 310 - 320 - 330 - 340 - 350 - 360 - 370 - 380 - 390
400 - 410 - 420 - 430 - 440 - 450 - 460 - 470 - 480 - 490 - 500 - 510 - 520
530 - 540 - 550 - 560 - 570 - 580 - 590 - 600 - 610 - 620 - 630 - 640 - 650
660 - 670 - 680 - 690 - 700 - 710 - 720 - 730 - 740 - 750 - 760 - 770 - 780
790 - 800 - 810 - 820 - 830 - 840 - 850 - 860 - 870 - 880 - 890 - 900 - 910
920 - 930 - 940 - 950 - 960 - 970 - 980 - 990 - 1000

Reg Reps: 10 - 20 - 30 - 40 - 50 - 60 - 70 - 80 - 90 - 100 - 110 - 120 - 130
140 - 150 - 160 - 170 - 180 - 190 - 200 - 210 - 220 - 230 - 240 - 250 - 260
270 - 280 - 290 - 300 - 310 - 320 - 330 - 340 - 350 - 360 - 370 - 380 - 390
400 - 410 - 420 - 430 - 440 - 450 - 460 - 470 - 480 - 490 - 500 - 510 - 520
530 - 540 - 550 - 560 - 570 - 580 - 590 - 600 - 610 - 620 - 630 - 640 - 650
660 - 670 - 680 - 690 - 700 - 710 - 720 - 730 - 740 - 750 - 760 - 770 - 780
790 - 800 - 810 - 820 - 830 - 840 - 850 - 860 - 870 - 880 - 890 - 900 - 910
920 - 930 - 940 - 950 - 960 - 970 - 980 - 990 - 1000

Notes: ____

____

____

____

Reg Reps: 10 - 20 - 30 - 40 - 50 - 60 - 70 - 80 - 90 - 100 - 110 - 120 - 130
140 - 150 - 160 - 170 - 180 - 190 - 200 - 210 - 220 - 230 - 240 - 250 - 260
270 - 280 - 290 - 300 - 310 - 320 - 330 - 340 - 350 - 360 - 370 - 380 - 390
400 - 410 - 420 - 430 - 440 - 450 - 460 - 470 - 480 - 490 - 500 - 510 - 520
530 - 540 - 550 - 560 - 570 - 580 - 590 - 600 - 610 - 620 - 630 - 640 - 650
660 - 670 - 680 - 690 - 700 - 710 - 720 - 730 - 740 - 750 - 760 - 770 - 780
790 - 800 - 810 - 820 - 830 - 840 - 850 - 860 - 870 - 880 - 890 - 900 - 910
920 - 930 - 940 - 950 - 960 - 970 - 980 - 990 - 1000

Notes: ____

____

____

____

Reg Reps: 10 - 20 - 30 - 40 - 50 - 60 - 70 - 80 - 90 - 100 - 110 - 120 - 130
140 - 150 - 160 - 170 - 180 - 190 - 200 - 210 - 220 - 230 - 240 - 250 - 260
270 - 280 - 290 - 300 - 310 - 320 - 330 - 340 - 350 - 360 - 370 - 380 - 390
400 - 410 - 420 - 430 - 440 - 450 - 460 - 470 - 480 - 490 - 500 - 510 - 520
530 - 540 - 550 - 560 - 570 - 580 - 590 - 600 - 610 - 620 - 630 - 640 - 650
660 - 670 - 680 - 690 - 700 - 710 - 720 - 730 - 740 - 750 - 760 - 770 - 780
790 - 800 - 810 - 820 - 830 - 840 - 850 - 860 - 870 - 880 - 890 - 900 - 910
920 - 930 - 940 - 950 - 960 - 970 - 980 - 990 - 1000

Notes: ____

____

____

____

Reg Reps: 10 - 20 - 30 - 40 - 50 - 60 - 70 - 80 - 90 - 100 - 110 - 120 - 130
140 - 150 - 160 - 170 - 180 - 190 - 200 - 210 - 220 - 230 - 240 - 250 - 260
270 - 280 - 290 - 300 - 310 - 320 - 330 - 340 - 350 - 360 - 370 - 380 - 390
400 - 410 - 420 - 430 - 440 - 450 - 460 - 470 - 480 - 490 - 500 - 510 - 520
530 - 540 - 550 - 560 - 570 - 580 - 590 - 600 - 610 - 620 - 630 - 640 - 650
660 - 670 - 680 - 690 - 700 - 710 - 720 - 730 - 740 - 750 - 760 - 770 - 780
790 - 800 - 810 - 820 - 830 - 840 - 850 - 860 - 870 - 880 - 890 - 900 - 910
920 - 930 - 940 - 950 - 960 - 970 - 980 - 990 - 1000

Notes: ____

____

____

____

Reg Reps: 10 - 20 - 30 - 40 - 50 - 60 - 70 - 80 - 90 - 100 - 110 - 120 - 130
140 - 150 - 160 - 170 - 180 - 190 - 200 - 210 - 220 - 230 - 240 - 250 - 260
270 - 280 - 290 - 300 - 310 - 320 - 330 - 340 - 350 - 360 - 370 - 380 - 390
400 - 410 - 420 - 430 - 440 - 450 - 460 - 470 - 480 - 490 - 500 - 510 - 520
530 - 540 - 550 - 560 - 570 - 580 - 590 - 600 - 610 - 620 - 630 - 640 - 650
660 - 670 - 680 - 690 - 700 - 710 - 720 - 730 - 740 - 750 - 760 - 770 - 780
790 - 800 - 810 - 820 - 830 - 840 - 850 - 860 - 870 - 880 - 890 - 900 - 910
920 - 930 - 940 - 950 - 960 - 970 - 980 - 990 - 1000

Notes: ____

____

____

______________________________ **End Date:** ____________

Kick: ______________________________ **Start Date:** __________

Kicking Leg: Left - Right - Alt. **Kicking Height: Low - Mid - High**

Notes: ____

____

____

Slow Reps: 10 - 20 - 30 - 40 - 50 - 60 - 70 - 80 - 90 - 100 - 110 - 120 - 130
140 - 150 - 160 - 170 - 180 - 190 - 200 - 210 - 220 - 230 - 240 - 250 - 260
270 - 280 - 290 - 300 - 310 - 320 - 330 - 340 - 350 - 360 - 370 - 380 - 390
400 - 410 - 420 - 430 - 440 - 450 - 460 - 470 - 480 - 490 - 500 - 510 - 520
530 - 540 - 550 - 560 - 570 - 580 - 590 - 600 - 610 - 620 - 630 - 640 - 650
660 - 670 - 680 - 690 - 700 - 710 - 720 - 730 - 740 - 750 - 760 - 770 - 780
790 - 800 - 810 - 820 - 830 - 840 - 850 - 860 - 870 - 880 - 890 - 900 - 910
920 - 930 - 940 - 950 - 960 - 970 - 980 - 990 - 1000

Reg Reps: 10 - 20 - 30 - 40 - 50 - 60 - 70 - 80 - 90 - 100 - 110 - 120 - 130
140 - 150 - 160 - 170 - 180 - 190 - 200 - 210 - 220 - 230 - 240 - 250 - 260
270 - 280 - 290 - 300 - 310 - 320 - 330 - 340 - 350 - 360 - 370 - 380 - 390
400 - 410 - 420 - 430 - 440 - 450 - 460 - 470 - 480 - 490 - 500 - 510 - 520
530 - 540 - 550 - 560 - 570 - 580 - 590 - 600 - 610 - 620 - 630 - 640 - 650
660 - 670 - 680 - 690 - 700 - 710 - 720 - 730 - 740 - 750 - 760 - 770 - 780
790 - 800 - 810 - 820 - 830 - 840 - 850 - 860 - 870 - 880 - 890 - 900 - 910
920 - 930 - 940 - 950 - 960 - 970 - 980 - 990 - 1000

Notes: ____

____

____

____

Reg Reps: 10 - 20 - 30 - 40 - 50 - 60 - 70 - 80 - 90 - 100 - 110 - 120 - 130
140 - 150 - 160 - 170 - 180 - 190 - 200 - 210 - 220 - 230 - 240 - 250 - 260
270 - 280 - 290 - 300 - 310 - 320 - 330 - 340 - 350 - 360 - 370 - 380 - 390
400 - 410 - 420 - 430 - 440 - 450 - 460 - 470 - 480 - 490 - 500 - 510 - 520
530 - 540 - 550 - 560 - 570 - 580 - 590 - 600 - 610 - 620 - 630 - 640 - 650
660 - 670 - 680 - 690 - 700 - 710 - 720 - 730 - 740 - 750 - 760 - 770 - 780
790 - 800 - 810 - 820 - 830 - 840 - 850 - 860 - 870 - 880 - 890 - 900 - 910
920 - 930 - 940 - 950 - 960 - 970 - 980 - 990 - 1000

Notes: ____

____

____

____

Reg Reps: 10 - 20 - 30 - 40 - 50 - 60 - 70 - 80 - 90 - 100 - 110 - 120 - 130
140 - 150 - 160 - 170 - 180 - 190 - 200 - 210 - 220 - 230 - 240 - 250 - 260
270 - 280 - 290 - 300 - 310 - 320 - 330 - 340 - 350 - 360 - 370 - 380 - 390
400 - 410 - 420 - 430 - 440 - 450 - 460 - 470 - 480 - 490 - 500 - 510 - 520
530 - 540 - 550 - 560 - 570 - 580 - 590 - 600 - 610 - 620 - 630 - 640 - 650
660 - 670 - 680 - 690 - 700 - 710 - 720 - 730 - 740 - 750 - 760 - 770 - 780
790 - 800 - 810 - 820 - 830 - 840 - 850 - 860 - 870 - 880 - 890 - 900 - 910
920 - 930 - 940 - 950 - 960 - 970 - 980 - 990 - 1000

Notes: ______________________________

____

____

____

Reg Reps: 10 - 20 - 30 - 40 - 50 - 60 - 70 - 80 - 90 - 100 - 110 - 120 - 130
140 - 150 - 160 - 170 - 180 - 190 - 200 - 210 - 220 - 230 - 240 - 250 - 260
270 - 280 - 290 - 300 - 310 - 320 - 330 - 340 - 350 - 360 - 370 - 380 - 390
400 - 410 - 420 - 430 - 440 - 450 - 460 - 470 - 480 - 490 - 500 - 510 - 520
530 - 540 - 550 - 560 - 570 - 580 - 590 - 600 - 610 - 620 - 630 - 640 - 650
660 - 670 - 680 - 690 - 700 - 710 - 720 - 730 - 740 - 750 - 760 - 770 - 780
790 - 800 - 810 - 820 - 830 - 840 - 850 - 860 - 870 - 880 - 890 - 900 - 910
920 - 930 - 940 - 950 - 960 - 970 - 980 - 990 - 1000

Notes: ______________________________

____

____

____

Reg Reps: 10 - 20 - 30 - 40 - 50 - 60 - 70 - 80 - 90 - 100 - 110 - 120 - 130
140 - 150 - 160 - 170 - 180 - 190 - 200 - 210 - 220 - 230 - 240 - 250 - 260
270 - 280 - 290 - 300 - 310 - 320 - 330 - 340 - 350 - 360 - 370 - 380 - 390
400 - 410 - 420 - 430 - 440 - 450 - 460 - 470 - 480 - 490 - 500 - 510 - 520
530 - 540 - 550 - 560 - 570 - 580 - 590 - 600 - 610 - 620 - 630 - 640 - 650
660 - 670 - 680 - 690 - 700 - 710 - 720 - 730 - 740 - 750 - 760 - 770 - 780
790 - 800 - 810 - 820 - 830 - 840 - 850 - 860 - 870 - 880 - 890 - 900 - 910
920 - 930 - 940 - 950 - 960 - 970 - 980 - 990 - 1000

Notes: ______________________________

____

____

______________________________ **End Date:** ____________

Kick:____________________________________ **Start Date:** __________

Kicking Leg: Left - Right - Alt. **Kicking Height: Low - Mid - High**

Notes:____

____

____

Slow Reps: 10 - 20 - 30 - 40 - 50 - 60 - 70 - 80 - 90 - 100 - 110 - 120 - 130
140 - 150 - 160 - 170 - 180 - 190 - 200 - 210 - 220 - 230 - 240 - 250 - 260
270 - 280 - 290 - 300 - 310 - 320 - 330 - 340 - 350 - 360 - 370 - 380 - 390
400 - 410 - 420 - 430 - 440 - 450 - 460 - 470 - 480 - 490 - 500 - 510 - 520
530 - 540 - 550 - 560 - 570 - 580 - 590 - 600 - 610 - 620 - 630 - 640 - 650
660 - 670 - 680 - 690 - 700 - 710 - 720 - 730 - 740 - 750 - 760 - 770 - 780
790 - 800 - 810 - 820 - 830 - 840 - 850 - 860 - 870 - 880 - 890 - 900 - 910
920 - 930 - 940 - 950 - 960 - 970 - 980 - 990 - 1000

Reg Reps: 10 - 20 - 30 - 40 - 50 - 60 - 70 - 80 - 90 - 100 - 110 - 120 - 130
140 - 150 - 160 - 170 - 180 - 190 - 200 - 210 - 220 - 230 - 240 - 250 - 260
270 - 280 - 290 - 300 - 310 - 320 - 330 - 340 - 350 - 360 - 370 - 380 - 390
400 - 410 - 420 - 430 - 440 - 450 - 460 - 470 - 480 - 490 - 500 - 510 - 520
530 - 540 - 550 - 560 - 570 - 580 - 590 - 600 - 610 - 620 - 630 - 640 - 650
660 - 670 - 680 - 690 - 700 - 710 - 720 - 730 - 740 - 750 - 760 - 770 - 780
790 - 800 - 810 - 820 - 830 - 840 - 850 - 860 - 870 - 880 - 890 - 900 - 910
920 - 930 - 940 - 950 - 960 - 970 - 980 - 990 - 1000

Notes: ____

____

____

____

Reg Reps: 10 - 20 - 30 - 40 - 50 - 60 - 70 - 80 - 90 - 100 - 110 - 120 - 130
140 - 150 - 160 - 170 - 180 - 190 - 200 - 210 - 220 - 230 - 240 - 250 - 260
270 - 280 - 290 - 300 - 310 - 320 - 330 - 340 - 350 - 360 - 370 - 380 - 390
400 - 410 - 420 - 430 - 440 - 450 - 460 - 470 - 480 - 490 - 500 - 510 - 520
530 - 540 - 550 - 560 - 570 - 580 - 590 - 600 - 610 - 620 - 630 - 640 - 650
660 - 670 - 680 - 690 - 700 - 710 - 720 - 730 - 740 - 750 - 760 - 770 - 780
790 - 800 - 810 - 820 - 830 - 840 - 850 - 860 - 870 - 880 - 890 - 900 - 910
920 - 930 - 940 - 950 - 960 - 970 - 980 - 990 - 1000

Notes: ____

____

____

____

Reg Reps: 10 - 20 - 30 - 40 - 50 - 60 - 70 - 80 - 90 - 100 - 110 - 120 - 130 140 - 150 - 160 - 170 - 180 - 190 - 200 - 210 - 220 - 230 - 240 - 250 - 260 270 - 280 - 290 - 300 - 310 - 320 - 330 - 340 - 350 - 360 - 370 - 380 - 390 400 - 410 - 420 - 430 - 440 - 450 - 460 - 470 - 480 - 490 - 500 - 510 - 520 530 - 540 - 550 - 560 - 570 - 580 - 590 - 600 - 610 - 620 - 630 - 640 - 650 660 - 670 - 680 - 690 - 700 - 710 - 720 - 730 - 740 - 750 - 760 - 770 - 780 790 - 800 - 810 - 820 - 830 - 840 - 850 - 860 - 870 - 880 - 890 - 900 - 910 920 - 930 - 940 - 950 - 960 - 970 - 980 - 990 - 1000

Notes: ____

____

____

____

Reg Reps: 10 - 20 - 30 - 40 - 50 - 60 - 70 - 80 - 90 - 100 - 110 - 120 - 130 140 - 150 - 160 - 170 - 180 - 190 - 200 - 210 - 220 - 230 - 240 - 250 - 260 270 - 280 - 290 - 300 - 310 - 320 - 330 - 340 - 350 - 360 - 370 - 380 - 390 400 - 410 - 420 - 430 - 440 - 450 - 460 - 470 - 480 - 490 - 500 - 510 - 520 530 - 540 - 550 - 560 - 570 - 580 - 590 - 600 - 610 - 620 - 630 - 640 - 650 660 - 670 - 680 - 690 - 700 - 710 - 720 - 730 - 740 - 750 - 760 - 770 - 780 790 - 800 - 810 - 820 - 830 - 840 - 850 - 860 - 870 - 880 - 890 - 900 - 910 920 - 930 - 940 - 950 - 960 - 970 - 980 - 990 - 1000

Notes: ____

____

____

____

Reg Reps: 10 - 20 - 30 - 40 - 50 - 60 - 70 - 80 - 90 - 100 - 110 - 120 - 130 140 - 150 - 160 - 170 - 180 - 190 - 200 - 210 - 220 - 230 - 240 - 250 - 260 270 - 280 - 290 - 300 - 310 - 320 - 330 - 340 - 350 - 360 - 370 - 380 - 390 400 - 410 - 420 - 430 - 440 - 450 - 460 - 470 - 480 - 490 - 500 - 510 - 520 530 - 540 - 550 - 560 - 570 - 580 - 590 - 600 - 610 - 620 - 630 - 640 - 650 660 - 670 - 680 - 690 - 700 - 710 - 720 - 730 - 740 - 750 - 760 - 770 - 780 790 - 800 - 810 - 820 - 830 - 840 - 850 - 860 - 870 - 880 - 890 - 900 - 910 920 - 930 - 940 - 950 - 960 - 970 - 980 - 990 - 1000

Notes: ____

____

____

______________________________ **End Date:** ____________

Kick: ______________________________ **Start Date:** ________
Kicking Leg: Left - Right - Alt. **Kicking Height: Low - Mid - High**
Notes: ____
____
____

Slow Reps: 10 - 20 - 30 - 40 - 50 - 60 - 70 - 80 - 90 - 100 - 110 - 120 - 130
140 - 150 - 160 - 170 - 180 - 190 - 200 - 210 - 220 - 230 - 240 - 250 - 260
270 - 280 - 290 - 300 - 310 - 320 - 330 - 340 - 350 - 360 - 370 - 380 - 390
400 - 410 - 420 - 430 - 440 - 450 - 460 - 470 - 480 - 490 - 500 - 510 - 520
530 - 540 - 550 - 560 - 570 - 580 - 590 - 600 - 610 - 620 - 630 - 640 - 650
660 - 670 - 680 - 690 - 700 - 710 - 720 - 730 - 740 - 750 - 760 - 770 - 780
790 - 800 - 810 - 820 - 830 - 840 - 850 - 860 - 870 - 880 - 890 - 900 - 910
920 - 930 - 940 - 950 - 960 - 970 - 980 - 990 - 1000

Reg Reps: 10 - 20 - 30 - 40 - 50 - 60 - 70 - 80 - 90 - 100 - 110 - 120 - 130
140 - 150 - 160 - 170 - 180 - 190 - 200 - 210 - 220 - 230 - 240 - 250 - 260
270 - 280 - 290 - 300 - 310 - 320 - 330 - 340 - 350 - 360 - 370 - 380 - 390
400 - 410 - 420 - 430 - 440 - 450 - 460 - 470 - 480 - 490 - 500 - 510 - 520
530 - 540 - 550 - 560 - 570 - 580 - 590 - 600 - 610 - 620 - 630 - 640 - 650
660 - 670 - 680 - 690 - 700 - 710 - 720 - 730 - 740 - 750 - 760 - 770 - 780
790 - 800 - 810 - 820 - 830 - 840 - 850 - 860 - 870 - 880 - 890 - 900 - 910
920 - 930 - 940 - 950 - 960 - 970 - 980 - 990 - 1000
Notes: ____
____
____
____

Reg Reps: 10 - 20 - 30 - 40 - 50 - 60 - 70 - 80 - 90 - 100 - 110 - 120 - 130
140 - 150 - 160 - 170 - 180 - 190 - 200 - 210 - 220 - 230 - 240 - 250 - 260
270 - 280 - 290 - 300 - 310 - 320 - 330 - 340 - 350 - 360 - 370 - 380 - 390
400 - 410 - 420 - 430 - 440 - 450 - 460 - 470 - 480 - 490 - 500 - 510 - 520
530 - 540 - 550 - 560 - 570 - 580 - 590 - 600 - 610 - 620 - 630 - 640 - 650
660 - 670 - 680 - 690 - 700 - 710 - 720 - 730 - 740 - 750 - 760 - 770 - 780
790 - 800 - 810 - 820 - 830 - 840 - 850 - 860 - 870 - 880 - 890 - 900 - 910
920 - 930 - 940 - 950 - 960 - 970 - 980 - 990 - 1000
Notes: ____
____
____
____

Reg Reps: 10 - 20 - 30 - 40 - 50 - 60 - 70 - 80 - 90 - 100 - 110 - 120 - 130
140 - 150 - 160 - 170 - 180 - 190 - 200 - 210 - 220 - 230 - 240 - 250 - 260
270 - 280 - 290 - 300 - 310 - 320 - 330 - 340 - 350 - 360 - 370 - 380 - 390
400 - 410 - 420 - 430 - 440 - 450 - 460 - 470 - 480 - 490 - 500 - 510 - 520
530 - 540 - 550 - 560 - 570 - 580 - 590 - 600 - 610 - 620 - 630 - 640 - 650
660 - 670 - 680 - 690 - 700 - 710 - 720 - 730 - 740 - 750 - 760 - 770 - 780
790 - 800 - 810 - 820 - 830 - 840 - 850 - 860 - 870 - 880 - 890 - 900 - 910
920 - 930 - 940 - 950 - 960 - 970 - 980 - 990 - 1000

Notes: ____

____

____

____

Reg Reps: 10 - 20 - 30 - 40 - 50 - 60 - 70 - 80 - 90 - 100 - 110 - 120 - 130
140 - 150 - 160 - 170 - 180 - 190 - 200 - 210 - 220 - 230 - 240 - 250 - 260
270 - 280 - 290 - 300 - 310 - 320 - 330 - 340 - 350 - 360 - 370 - 380 - 390
400 - 410 - 420 - 430 - 440 - 450 - 460 - 470 - 480 - 490 - 500 - 510 - 520
530 - 540 - 550 - 560 - 570 - 580 - 590 - 600 - 610 - 620 - 630 - 640 - 650
660 - 670 - 680 - 690 - 700 - 710 - 720 - 730 - 740 - 750 - 760 - 770 - 780
790 - 800 - 810 - 820 - 830 - 840 - 850 - 860 - 870 - 880 - 890 - 900 - 910
920 - 930 - 940 - 950 - 960 - 970 - 980 - 990 - 1000

Notes: ____

____

____

____

Reg Reps: 10 - 20 - 30 - 40 - 50 - 60 - 70 - 80 - 90 - 100 - 110 - 120 - 130
140 - 150 - 160 - 170 - 180 - 190 - 200 - 210 - 220 - 230 - 240 - 250 - 260
270 - 280 - 290 - 300 - 310 - 320 - 330 - 340 - 350 - 360 - 370 - 380 - 390
400 - 410 - 420 - 430 - 440 - 450 - 460 - 470 - 480 - 490 - 500 - 510 - 520
530 - 540 - 550 - 560 - 570 - 580 - 590 - 600 - 610 - 620 - 630 - 640 - 650
660 - 670 - 680 - 690 - 700 - 710 - 720 - 730 - 740 - 750 - 760 - 770 - 780
790 - 800 - 810 - 820 - 830 - 840 - 850 - 860 - 870 - 880 - 890 - 900 - 910
920 - 930 - 940 - 950 - 960 - 970 - 980 - 990 - 1000

Notes: ____

____

____

______________________________ **End Date:** ____________

Kick: ____________________ **Start Date:** ________

Kicking Leg: Left - Right - Alt. **Kicking Height: Low - Mid - High**

Notes: ____________________

____

____

Slow Reps: 10 - 20 - 30 - 40 - 50 - 60 - 70 - 80 - 90 - 100 - 110 - 120 - 130
140 - 150 - 160 - 170 - 180 - 190 - 200 - 210 - 220 - 230 - 240 - 250 - 260
270 - 280 - 290 - 300 - 310 - 320 - 330 - 340 - 350 - 360 - 370 - 380 - 390
400 - 410 - 420 - 430 - 440 - 450 - 460 - 470 - 480 - 490 - 500 - 510 - 520
530 - 540 - 550 - 560 - 570 - 580 - 590 - 600 - 610 - 620 - 630 - 640 - 650
660 - 670 - 680 - 690 - 700 - 710 - 720 - 730 - 740 - 750 - 760 - 770 - 780
790 - 800 - 810 - 820 - 830 - 840 - 850 - 860 - 870 - 880 - 890 - 900 - 910
920 - 930 - 940 - 950 - 960 - 970 - 980 - 990 - 1000

Reg Reps: 10 - 20 - 30 - 40 - 50 - 60 - 70 - 80 - 90 - 100 - 110 - 120 - 130
140 - 150 - 160 - 170 - 180 - 190 - 200 - 210 - 220 - 230 - 240 - 250 - 260
270 - 280 - 290 - 300 - 310 - 320 - 330 - 340 - 350 - 360 - 370 - 380 - 390
400 - 410 - 420 - 430 - 440 - 450 - 460 - 470 - 480 - 490 - 500 - 510 - 520
530 - 540 - 550 - 560 - 570 - 580 - 590 - 600 - 610 - 620 - 630 - 640 - 650
660 - 670 - 680 - 690 - 700 - 710 - 720 - 730 - 740 - 750 - 760 - 770 - 780
790 - 800 - 810 - 820 - 830 - 840 - 850 - 860 - 870 - 880 - 890 - 900 - 910
920 - 930 - 940 - 950 - 960 - 970 - 980 - 990 - 1000

Notes: ____________________

____

____

____

Reg Reps: 10 - 20 - 30 - 40 - 50 - 60 - 70 - 80 - 90 - 100 - 110 - 120 - 130
140 - 150 - 160 - 170 - 180 - 190 - 200 - 210 - 220 - 230 - 240 - 250 - 260
270 - 280 - 290 - 300 - 310 - 320 - 330 - 340 - 350 - 360 - 370 - 380 - 390
400 - 410 - 420 - 430 - 440 - 450 - 460 - 470 - 480 - 490 - 500 - 510 - 520
530 - 540 - 550 - 560 - 570 - 580 - 590 - 600 - 610 - 620 - 630 - 640 - 650
660 - 670 - 680 - 690 - 700 - 710 - 720 - 730 - 740 - 750 - 760 - 770 - 780
790 - 800 - 810 - 820 - 830 - 840 - 850 - 860 - 870 - 880 - 890 - 900 - 910
920 - 930 - 940 - 950 - 960 - 970 - 980 - 990 - 1000

Notes: ____________________

____

____

____

Reg Reps: 10 - 20 - 30 - 40 - 50 - 60 - 70 - 80 - 90 - 100 - 110 - 120 - 130 - 140 - 150 - 160 - 170 - 180 - 190 - 200 - 210 - 220 - 230 - 240 - 250 - 260 - 270 - 280 - 290 - 300 - 310 - 320 - 330 - 340 - 350 - 360 - 370 - 380 - 390 - 400 - 410 - 420 - 430 - 440 - 450 - 460 - 470 - 480 - 490 - 500 - 510 - 520 - 530 - 540 - 550 - 560 - 570 - 580 - 590 - 600 - 610 - 620 - 630 - 640 - 650 - 660 - 670 - 680 - 690 - 700 - 710 - 720 - 730 - 740 - 750 - 760 - 770 - 780 - 790 - 800 - 810 - 820 - 830 - 840 - 850 - 860 - 870 - 880 - 890 - 900 - 910 - 920 - 930 - 940 - 950 - 960 - 970 - 980 - 990 - 1000

Notes: ____

____

____

____

Reg Reps: 10 - 20 - 30 - 40 - 50 - 60 - 70 - 80 - 90 - 100 - 110 - 120 - 130 - 140 - 150 - 160 - 170 - 180 - 190 - 200 - 210 - 220 - 230 - 240 - 250 - 260 - 270 - 280 - 290 - 300 - 310 - 320 - 330 - 340 - 350 - 360 - 370 - 380 - 390 - 400 - 410 - 420 - 430 - 440 - 450 - 460 - 470 - 480 - 490 - 500 - 510 - 520 - 530 - 540 - 550 - 560 - 570 - 580 - 590 - 600 - 610 - 620 - 630 - 640 - 650 - 660 - 670 - 680 - 690 - 700 - 710 - 720 - 730 - 740 - 750 - 760 - 770 - 780 - 790 - 800 - 810 - 820 - 830 - 840 - 850 - 860 - 870 - 880 - 890 - 900 - 910 - 920 - 930 - 940 - 950 - 960 - 970 - 980 - 990 - 1000

Notes: ____

____

____

____

Reg Reps: 10 - 20 - 30 - 40 - 50 - 60 - 70 - 80 - 90 - 100 - 110 - 120 - 130 - 140 - 150 - 160 - 170 - 180 - 190 - 200 - 210 - 220 - 230 - 240 - 250 - 260 - 270 - 280 - 290 - 300 - 310 - 320 - 330 - 340 - 350 - 360 - 370 - 380 - 390 - 400 - 410 - 420 - 430 - 440 - 450 - 460 - 470 - 480 - 490 - 500 - 510 - 520 - 530 - 540 - 550 - 560 - 570 - 580 - 590 - 600 - 610 - 620 - 630 - 640 - 650 - 660 - 670 - 680 - 690 - 700 - 710 - 720 - 730 - 740 - 750 - 760 - 770 - 780 - 790 - 800 - 810 - 820 - 830 - 840 - 850 - 860 - 870 - 880 - 890 - 900 - 910 - 920 - 930 - 940 - 950 - 960 - 970 - 980 - 990 - 1000

Notes: ____

____

____

______________________________ **End Date:** ____________

Kick:________________________________ **Start Date:** ________

Kicking Leg: Left - Right - Alt. **Kicking Height: Low - Mid - High**

Notes:____

Slow Reps: 10 - 20 - 30 - 40 - 50 - 60 - 70 - 80 - 90 - 100 - 110 - 120 - 130
140 - 150 - 160 - 170 - 180 - 190 - 200 - 210 - 220 - 230 - 240 - 250 - 260
270 - 280 - 290 - 300 - 310 - 320 - 330 - 340 - 350 - 360 - 370 - 380 - 390
400 - 410 - 420 - 430 - 440 - 450 - 460 - 470 - 480 - 490 - 500 - 510 - 520
530 - 540 - 550 - 560 - 570 - 580 - 590 - 600 - 610 - 620 - 630 - 640 - 650
660 - 670 - 680 - 690 - 700 - 710 - 720 - 730 - 740 - 750 - 760 - 770 - 780
790 - 800 - 810 - 820 - 830 - 840 - 850 - 860 - 870 - 880 - 890 - 900 - 910
920 - 930 - 940 - 950 - 960 - 970 - 980 - 990 - 1000

Reg Reps: 10 - 20 - 30 - 40 - 50 - 60 - 70 - 80 - 90 - 100 - 110 - 120 - 130
140 - 150 - 160 - 170 - 180 - 190 - 200 - 210 - 220 - 230 - 240 - 250 - 260
270 - 280 - 290 - 300 - 310 - 320 - 330 - 340 - 350 - 360 - 370 - 380 - 390
400 - 410 - 420 - 430 - 440 - 450 - 460 - 470 - 480 - 490 - 500 - 510 - 520
530 - 540 - 550 - 560 - 570 - 580 - 590 - 600 - 610 - 620 - 630 - 640 - 650
660 - 670 - 680 - 690 - 700 - 710 - 720 - 730 - 740 - 750 - 760 - 770 - 780
790 - 800 - 810 - 820 - 830 - 840 - 850 - 860 - 870 - 880 - 890 - 900 - 910
920 - 930 - 940 - 950 - 960 - 970 - 980 - 990 - 1000

Notes: ____

Reg Reps: 10 - 20 - 30 - 40 - 50 - 60 - 70 - 80 - 90 - 100 - 110 - 120 - 130
140 - 150 - 160 - 170 - 180 - 190 - 200 - 210 - 220 - 230 - 240 - 250 - 260
270 - 280 - 290 - 300 - 310 - 320 - 330 - 340 - 350 - 360 - 370 - 380 - 390
400 - 410 - 420 - 430 - 440 - 450 - 460 - 470 - 480 - 490 - 500 - 510 - 520
530 - 540 - 550 - 560 - 570 - 580 - 590 - 600 - 610 - 620 - 630 - 640 - 650
660 - 670 - 680 - 690 - 700 - 710 - 720 - 730 - 740 - 750 - 760 - 770 - 780
790 - 800 - 810 - 820 - 830 - 840 - 850 - 860 - 870 - 880 - 890 - 900 - 910
920 - 930 - 940 - 950 - 960 - 970 - 980 - 990 - 1000

Notes: ____

Reg Reps: 10 - 20 - 30 - 40 - 50 - 60 - 70 - 80 - 90 - 100 - 110 - 120 - 130
140 - 150 - 160 - 170 - 180 - 190 - 200 - 210 - 220 - 230 - 240 - 250 - 260
270 - 280 - 290 - 300 - 310 - 320 - 330 - 340 - 350 - 360 - 370 - 380 - 390
400 - 410 - 420 - 430 - 440 - 450 - 460 - 470 - 480 - 490 - 500 - 510 - 520
530 - 540 - 550 - 560 - 570 - 580 - 590 - 600 - 610 - 620 - 630 - 640 - 650
660 - 670 - 680 - 690 - 700 - 710 - 720 - 730 - 740 - 750 - 760 - 770 - 780
790 - 800 - 810 - 820 - 830 - 840 - 850 - 860 - 870 - 880 - 890 - 900 - 910
920 - 930 - 940 - 950 - 960 - 970 - 980 - 990 - 1000

Notes: ____

____

____

____

Reg Reps: 10 - 20 - 30 - 40 - 50 - 60 - 70 - 80 - 90 - 100 - 110 - 120 - 130
140 - 150 - 160 - 170 - 180 - 190 - 200 - 210 - 220 - 230 - 240 - 250 - 260
270 - 280 - 290 - 300 - 310 - 320 - 330 - 340 - 350 - 360 - 370 - 380 - 390
400 - 410 - 420 - 430 - 440 - 450 - 460 - 470 - 480 - 490 - 500 - 510 - 520
530 - 540 - 550 - 560 - 570 - 580 - 590 - 600 - 610 - 620 - 630 - 640 - 650
660 - 670 - 680 - 690 - 700 - 710 - 720 - 730 - 740 - 750 - 760 - 770 - 780
790 - 800 - 810 - 820 - 830 - 840 - 850 - 860 - 870 - 880 - 890 - 900 - 910
920 - 930 - 940 - 950 - 960 - 970 - 980 - 990 - 1000

Notes: ____

____

____

____

Reg Reps: 10 - 20 - 30 - 40 - 50 - 60 - 70 - 80 - 90 - 100 - 110 - 120 - 130
140 - 150 - 160 - 170 - 180 - 190 - 200 - 210 - 220 - 230 - 240 - 250 - 260
270 - 280 - 290 - 300 - 310 - 320 - 330 - 340 - 350 - 360 - 370 - 380 - 390
400 - 410 - 420 - 430 - 440 - 450 - 460 - 470 - 480 - 490 - 500 - 510 - 520
530 - 540 - 550 - 560 - 570 - 580 - 590 - 600 - 610 - 620 - 630 - 640 - 650
660 - 670 - 680 - 690 - 700 - 710 - 720 - 730 - 740 - 750 - 760 - 770 - 780
790 - 800 - 810 - 820 - 830 - 840 - 850 - 860 - 870 - 880 - 890 - 900 - 910
920 - 930 - 940 - 950 - 960 - 970 - 980 - 990 - 1000

Notes: ____

____

____

______________________________ **End Date:** __________

Kick:____ **Start Date:** __________

Kicking Leg: Left - Right - Alt. **Kicking Height: Low - Mid - High**

Notes:____

____

____

Slow Reps: 10 - 20 - 30 - 40 - 50 - 60 - 70 - 80 - 90 - 100 - 110 - 120 - 130
140 - 150 - 160 - 170 - 180 - 190 - 200 - 210 - 220 - 230 - 240 - 250 - 260
270 - 280 - 290 - 300 - 310 - 320 - 330 - 340 - 350 - 360 - 370 - 380 - 390
400 - 410 - 420 - 430 - 440 - 450 - 460 - 470 - 480 - 490 - 500 - 510 - 520
530 - 540 - 550 - 560 - 570 - 580 - 590 - 600 - 610 - 620 - 630 - 640 - 650
660 - 670 - 680 - 690 - 700 - 710 - 720 - 730 - 740 - 750 - 760 - 770 - 780
790 - 800 - 810 - 820 - 830 - 840 - 850 - 860 - 870 - 880 - 890 - 900 - 910
920 - 930 - 940 - 950 - 960 - 970 - 980 - 990 - 1000

Reg Reps: 10 - 20 - 30 - 40 - 50 - 60 - 70 - 80 - 90 - 100 - 110 - 120 - 130
140 - 150 - 160 - 170 - 180 - 190 - 200 - 210 - 220 - 230 - 240 - 250 - 260
270 - 280 - 290 - 300 - 310 - 320 - 330 - 340 - 350 - 360 - 370 - 380 - 390
400 - 410 - 420 - 430 - 440 - 450 - 460 - 470 - 480 - 490 - 500 - 510 - 520
530 - 540 - 550 - 560 - 570 - 580 - 590 - 600 - 610 - 620 - 630 - 640 - 650
660 - 670 - 680 - 690 - 700 - 710 - 720 - 730 - 740 - 750 - 760 - 770 - 780
790 - 800 - 810 - 820 - 830 - 840 - 850 - 860 - 870 - 880 - 890 - 900 - 910
920 - 930 - 940 - 950 - 960 - 970 - 980 - 990 - 1000

Notes: ____

____

____

____

Reg Reps: 10 - 20 - 30 - 40 - 50 - 60 - 70 - 80 - 90 - 100 - 110 - 120 - 130
140 - 150 - 160 - 170 - 180 - 190 - 200 - 210 - 220 - 230 - 240 - 250 - 260
270 - 280 - 290 - 300 - 310 - 320 - 330 - 340 - 350 - 360 - 370 - 380 - 390
400 - 410 - 420 - 430 - 440 - 450 - 460 - 470 - 480 - 490 - 500 - 510 - 520
530 - 540 - 550 - 560 - 570 - 580 - 590 - 600 - 610 - 620 - 630 - 640 - 650
660 - 670 - 680 - 690 - 700 - 710 - 720 - 730 - 740 - 750 - 760 - 770 - 780
790 - 800 - 810 - 820 - 830 - 840 - 850 - 860 - 870 - 880 - 890 - 900 - 910
920 - 930 - 940 - 950 - 960 - 970 - 980 - 990 - 1000

Notes: ____

____

____

____

Reg Reps: 10 - 20 - 30 - 40 - 50 - 60 - 70 - 80 - 90 - 100 - 110 - 120 - 130 - 140 - 150 - 160 - 170 - 180 - 190 - 200 - 210 - 220 - 230 - 240 - 250 - 260 - 270 - 280 - 290 - 300 - 310 - 320 - 330 - 340 - 350 - 360 - 370 - 380 - 390 - 400 - 410 - 420 - 430 - 440 - 450 - 460 - 470 - 480 - 490 - 500 - 510 - 520 - 530 - 540 - 550 - 560 - 570 - 580 - 590 - 600 - 610 - 620 - 630 - 640 - 650 - 660 - 670 - 680 - 690 - 700 - 710 - 720 - 730 - 740 - 750 - 760 - 770 - 780 - 790 - 800 - 810 - 820 - 830 - 840 - 850 - 860 - 870 - 880 - 890 - 900 - 910 - 920 - 930 - 940 - 950 - 960 - 970 - 980 - 990 - 1000

Notes: ______________________________

____

____

____

Reg Reps: 10 - 20 - 30 - 40 - 50 - 60 - 70 - 80 - 90 - 100 - 110 - 120 - 130 - 140 - 150 - 160 - 170 - 180 - 190 - 200 - 210 - 220 - 230 - 240 - 250 - 260 - 270 - 280 - 290 - 300 - 310 - 320 - 330 - 340 - 350 - 360 - 370 - 380 - 390 - 400 - 410 - 420 - 430 - 440 - 450 - 460 - 470 - 480 - 490 - 500 - 510 - 520 - 530 - 540 - 550 - 560 - 570 - 580 - 590 - 600 - 610 - 620 - 630 - 640 - 650 - 660 - 670 - 680 - 690 - 700 - 710 - 720 - 730 - 740 - 750 - 760 - 770 - 780 - 790 - 800 - 810 - 820 - 830 - 840 - 850 - 860 - 870 - 880 - 890 - 900 - 910 - 920 - 930 - 940 - 950 - 960 - 970 - 980 - 990 - 1000

Notes: ______________________________

____

____

____

Reg Reps: 10 - 20 - 30 - 40 - 50 - 60 - 70 - 80 - 90 - 100 - 110 - 120 - 130 - 140 - 150 - 160 - 170 - 180 - 190 - 200 - 210 - 220 - 230 - 240 - 250 - 260 - 270 - 280 - 290 - 300 - 310 - 320 - 330 - 340 - 350 - 360 - 370 - 380 - 390 - 400 - 410 - 420 - 430 - 440 - 450 - 460 - 470 - 480 - 490 - 500 - 510 - 520 - 530 - 540 - 550 - 560 - 570 - 580 - 590 - 600 - 610 - 620 - 630 - 640 - 650 - 660 - 670 - 680 - 690 - 700 - 710 - 720 - 730 - 740 - 750 - 760 - 770 - 780 - 790 - 800 - 810 - 820 - 830 - 840 - 850 - 860 - 870 - 880 - 890 - 900 - 910 - 920 - 930 - 940 - 950 - 960 - 970 - 980 - 990 - 1000

Notes: ______________________________

____

____

______________________________ **End Date:** ____________

Kick: ______________________________ **Start Date:** ________

Kicking Leg: Left - Right - Alt. **Kicking Height: Low - Mid - High**

Notes: ____

____

____

Slow Reps: 10 - 20 - 30 - 40 - 50 - 60 - 70 - 80 - 90 - 100 - 110 - 120 - 130
140 - 150 - 160 - 170 - 180 - 190 - 200 - 210 - 220 - 230 - 240 - 250 - 260
270 - 280 - 290 - 300 - 310 - 320 - 330 - 340 - 350 - 360 - 370 - 380 - 390
400 - 410 - 420 - 430 - 440 - 450 - 460 - 470 - 480 - 490 - 500 - 510 - 520
530 - 540 - 550 - 560 - 570 - 580 - 590 - 600 - 610 - 620 - 630 - 640 - 650
660 - 670 - 680 - 690 - 700 - 710 - 720 - 730 - 740 - 750 - 760 - 770 - 780
790 - 800 - 810 - 820 - 830 - 840 - 850 - 860 - 870 - 880 - 890 - 900 - 910
920 - 930 - 940 - 950 - 960 - 970 - 980 - 990 - 1000

Reg Reps: 10 - 20 - 30 - 40 - 50 - 60 - 70 - 80 - 90 - 100 - 110 - 120 - 130
140 - 150 - 160 - 170 - 180 - 190 - 200 - 210 - 220 - 230 - 240 - 250 - 260
270 - 280 - 290 - 300 - 310 - 320 - 330 - 340 - 350 - 360 - 370 - 380 - 390
400 - 410 - 420 - 430 - 440 - 450 - 460 - 470 - 480 - 490 - 500 - 510 - 520
530 - 540 - 550 - 560 - 570 - 580 - 590 - 600 - 610 - 620 - 630 - 640 - 650
660 - 670 - 680 - 690 - 700 - 710 - 720 - 730 - 740 - 750 - 760 - 770 - 780
790 - 800 - 810 - 820 - 830 - 840 - 850 - 860 - 870 - 880 - 890 - 900 - 910
920 - 930 - 940 - 950 - 960 - 970 - 980 - 990 - 1000

Notes: ____

____

____

____

Reg Reps: 10 - 20 - 30 - 40 - 50 - 60 - 70 - 80 - 90 - 100 - 110 - 120 - 130
140 - 150 - 160 - 170 - 180 - 190 - 200 - 210 - 220 - 230 - 240 - 250 - 260
270 - 280 - 290 - 300 - 310 - 320 - 330 - 340 - 350 - 360 - 370 - 380 - 390
400 - 410 - 420 - 430 - 440 - 450 - 460 - 470 - 480 - 490 - 500 - 510 - 520
530 - 540 - 550 - 560 - 570 - 580 - 590 - 600 - 610 - 620 - 630 - 640 - 650
660 - 670 - 680 - 690 - 700 - 710 - 720 - 730 - 740 - 750 - 760 - 770 - 780
790 - 800 - 810 - 820 - 830 - 840 - 850 - 860 - 870 - 880 - 890 - 900 - 910
920 - 930 - 940 - 950 - 960 - 970 - 980 - 990 - 1000

Notes: ____

____

____

____

Reg Reps: 10 - 20 - 30 - 40 - 50 - 60 - 70 - 80 - 90 - 100 - 110 - 120 - 130 - 140 - 150 - 160 - 170 - 180 - 190 - 200 - 210 - 220 - 230 - 240 - 250 - 260 - 270 - 280 - 290 - 300 - 310 - 320 - 330 - 340 - 350 - 360 - 370 - 380 - 390 - 400 - 410 - 420 - 430 - 440 - 450 - 460 - 470 - 480 - 490 - 500 - 510 - 520 - 530 - 540 - 550 - 560 - 570 - 580 - 590 - 600 - 610 - 620 - 630 - 640 - 650 - 660 - 670 - 680 - 690 - 700 - 710 - 720 - 730 - 740 - 750 - 760 - 770 - 780 - 790 - 800 - 810 - 820 - 830 - 840 - 850 - 860 - 870 - 880 - 890 - 900 - 910 - 920 - 930 - 940 - 950 - 960 - 970 - 980 - 990 - 1000

Notes: ______________________________

____

____

____

Reg Reps: 10 - 20 - 30 - 40 - 50 - 60 - 70 - 80 - 90 - 100 - 110 - 120 - 130 - 140 - 150 - 160 - 170 - 180 - 190 - 200 - 210 - 220 - 230 - 240 - 250 - 260 - 270 - 280 - 290 - 300 - 310 - 320 - 330 - 340 - 350 - 360 - 370 - 380 - 390 - 400 - 410 - 420 - 430 - 440 - 450 - 460 - 470 - 480 - 490 - 500 - 510 - 520 - 530 - 540 - 550 - 560 - 570 - 580 - 590 - 600 - 610 - 620 - 630 - 640 - 650 - 660 - 670 - 680 - 690 - 700 - 710 - 720 - 730 - 740 - 750 - 760 - 770 - 780 - 790 - 800 - 810 - 820 - 830 - 840 - 850 - 860 - 870 - 880 - 890 - 900 - 910 - 920 - 930 - 940 - 950 - 960 - 970 - 980 - 990 - 1000

Notes: ______________________________

____

____

____

Reg Reps: 10 - 20 - 30 - 40 - 50 - 60 - 70 - 80 - 90 - 100 - 110 - 120 - 130 - 140 - 150 - 160 - 170 - 180 - 190 - 200 - 210 - 220 - 230 - 240 - 250 - 260 - 270 - 280 - 290 - 300 - 310 - 320 - 330 - 340 - 350 - 360 - 370 - 380 - 390 - 400 - 410 - 420 - 430 - 440 - 450 - 460 - 470 - 480 - 490 - 500 - 510 - 520 - 530 - 540 - 550 - 560 - 570 - 580 - 590 - 600 - 610 - 620 - 630 - 640 - 650 - 660 - 670 - 680 - 690 - 700 - 710 - 720 - 730 - 740 - 750 - 760 - 770 - 780 - 790 - 800 - 810 - 820 - 830 - 840 - 850 - 860 - 870 - 880 - 890 - 900 - 910 - 920 - 930 - 940 - 950 - 960 - 970 - 980 - 990 - 1000

Notes: ______________________________

____

____

______________________________ **End Date:** ____________

Kick:____________________________________ **Start Date:** _________

Kicking Leg: Left - Right - Alt. **Kicking Height: Low - Mid - High**

Notes:____

____

____

Slow Reps: 10 - 20 - 30 - 40 - 50 - 60 - 70 - 80 - 90 - 100 - 110 - 120 - 130
140 - 150 - 160 - 170 - 180 - 190 - 200 - 210 - 220 - 230 - 240 - 250 - 260
270 - 280 - 290 - 300 - 310 - 320 - 330 - 340 - 350 - 360 - 370 - 380 - 390
400 - 410 - 420 - 430 - 440 - 450 - 460 - 470 - 480 - 490 - 500 - 510 - 520
530 - 540 - 550 - 560 - 570 - 580 - 590 - 600 - 610 - 620 - 630 - 640 - 650
660 - 670 - 680 - 690 - 700 - 710 - 720 - 730 - 740 - 750 - 760 - 770 - 780
790 - 800 - 810 - 820 - 830 - 840 - 850 - 860 - 870 - 880 - 890 - 900 - 910
920 - 930 - 940 - 950 - 960 - 970 - 980 - 990 - 1000

Reg Reps: 10 - 20 - 30 - 40 - 50 - 60 - 70 - 80 - 90 - 100 - 110 - 120 - 130
140 - 150 - 160 - 170 - 180 - 190 - 200 - 210 - 220 - 230 - 240 - 250 - 260
270 - 280 - 290 - 300 - 310 - 320 - 330 - 340 - 350 - 360 - 370 - 380 - 390
400 - 410 - 420 - 430 - 440 - 450 - 460 - 470 - 480 - 490 - 500 - 510 - 520
530 - 540 - 550 - 560 - 570 - 580 - 590 - 600 - 610 - 620 - 630 - 640 - 650
660 - 670 - 680 - 690 - 700 - 710 - 720 - 730 - 740 - 750 - 760 - 770 - 780
790 - 800 - 810 - 820 - 830 - 840 - 850 - 860 - 870 - 880 - 890 - 900 - 910
920 - 930 - 940 - 950 - 960 - 970 - 980 - 990 - 1000

Notes: ____

____

____

____

Reg Reps: 10 - 20 - 30 - 40 - 50 - 60 - 70 - 80 - 90 - 100 - 110 - 120 - 130
140 - 150 - 160 - 170 - 180 - 190 - 200 - 210 - 220 - 230 - 240 - 250 - 260
270 - 280 - 290 - 300 - 310 - 320 - 330 - 340 - 350 - 360 - 370 - 380 - 390
400 - 410 - 420 - 430 - 440 - 450 - 460 - 470 - 480 - 490 - 500 - 510 - 520
530 - 540 - 550 - 560 - 570 - 580 - 590 - 600 - 610 - 620 - 630 - 640 - 650
660 - 670 - 680 - 690 - 700 - 710 - 720 - 730 - 740 - 750 - 760 - 770 - 780
790 - 800 - 810 - 820 - 830 - 840 - 850 - 860 - 870 - 880 - 890 - 900 - 910
920 - 930 - 940 - 950 - 960 - 970 - 980 - 990 - 1000

Notes: ____

____

____

____

Reg Reps: 10 - 20 - 30 - 40 - 50 - 60 - 70 - 80 - 90 - 100 - 110 - 120 - 130
140 - 150 - 160 - 170 - 180 - 190 - 200 - 210 - 220 - 230 - 240 - 250 - 260
270 - 280 - 290 - 300 - 310 - 320 - 330 - 340 - 350 - 360 - 370 - 380 - 390
400 - 410 - 420 - 430 - 440 - 450 - 460 - 470 - 480 - 490 - 500 - 510 - 520
530 - 540 - 550 - 560 - 570 - 580 - 590 - 600 - 610 - 620 - 630 - 640 - 650
660 - 670 - 680 - 690 - 700 - 710 - 720 - 730 - 740 - 750 - 760 - 770 - 780
790 - 800 - 810 - 820 - 830 - 840 - 850 - 860 - 870 - 880 - 890 - 900 - 910
920 - 930 - 940 - 950 - 960 - 970 - 980 - 990 - 1000

Notes: ____

____

____

____

Reg Reps: 10 - 20 - 30 - 40 - 50 - 60 - 70 - 80 - 90 - 100 - 110 - 120 - 130
140 - 150 - 160 - 170 - 180 - 190 - 200 - 210 - 220 - 230 - 240 - 250 - 260
270 - 280 - 290 - 300 - 310 - 320 - 330 - 340 - 350 - 360 - 370 - 380 - 390
400 - 410 - 420 - 430 - 440 - 450 - 460 - 470 - 480 - 490 - 500 - 510 - 520
530 - 540 - 550 - 560 - 570 - 580 - 590 - 600 - 610 - 620 - 630 - 640 - 650
660 - 670 - 680 - 690 - 700 - 710 - 720 - 730 - 740 - 750 - 760 - 770 - 780
790 - 800 - 810 - 820 - 830 - 840 - 850 - 860 - 870 - 880 - 890 - 900 - 910
920 - 930 - 940 - 950 - 960 - 970 - 980 - 990 - 1000

Notes: ____

____

____

____

Reg Reps: 10 - 20 - 30 - 40 - 50 - 60 - 70 - 80 - 90 - 100 - 110 - 120 - 130
140 - 150 - 160 - 170 - 180 - 190 - 200 - 210 - 220 - 230 - 240 - 250 - 260
270 - 280 - 290 - 300 - 310 - 320 - 330 - 340 - 350 - 360 - 370 - 380 - 390
400 - 410 - 420 - 430 - 440 - 450 - 460 - 470 - 480 - 490 - 500 - 510 - 520
530 - 540 - 550 - 560 - 570 - 580 - 590 - 600 - 610 - 620 - 630 - 640 - 650
660 - 670 - 680 - 690 - 700 - 710 - 720 - 730 - 740 - 750 - 760 - 770 - 780
790 - 800 - 810 - 820 - 830 - 840 - 850 - 860 - 870 - 880 - 890 - 900 - 910
920 - 930 - 940 - 950 - 960 - 970 - 980 - 990 - 1000

Notes: ____

____

____

______________________________ **End Date:** ____________

Kick:________________________________ **Start Date:** ________

Kicking Leg: Left - Right - Alt. **Kicking Height: Low - Mid - High**

Notes:____

____

____

Slow Reps: 10 - 20 - 30 - 40 - 50 - 60 - 70 - 80 - 90 - 100 - 110 - 120 - 130
140 - 150 - 160 - 170 - 180 - 190 - 200 - 210 - 220 - 230 - 240 - 250 - 260
270 - 280 - 290 - 300 - 310 - 320 - 330 - 340 - 350 - 360 - 370 - 380 - 390
400 - 410 - 420 - 430 - 440 - 450 - 460 - 470 - 480 - 490 - 500 - 510 - 520
530 - 540 - 550 - 560 - 570 - 580 - 590 - 600 - 610 - 620 - 630 - 640 - 650
660 - 670 - 680 - 690 - 700 - 710 - 720 - 730 - 740 - 750 - 760 - 770 - 780
790 - 800 - 810 - 820 - 830 - 840 - 850 - 860 - 870 - 880 - 890 - 900 - 910
920 - 930 - 940 - 950 - 960 - 970 - 980 - 990 - 1000

Reg Reps: 10 - 20 - 30 - 40 - 50 - 60 - 70 - 80 - 90 - 100 - 110 - 120 - 130
140 - 150 - 160 - 170 - 180 - 190 - 200 - 210 - 220 - 230 - 240 - 250 - 260
270 - 280 - 290 - 300 - 310 - 320 - 330 - 340 - 350 - 360 - 370 - 380 - 390
400 - 410 - 420 - 430 - 440 - 450 - 460 - 470 - 480 - 490 - 500 - 510 - 520
530 - 540 - 550 - 560 - 570 - 580 - 590 - 600 - 610 - 620 - 630 - 640 - 650
660 - 670 - 680 - 690 - 700 - 710 - 720 - 730 - 740 - 750 - 760 - 770 - 780
790 - 800 - 810 - 820 - 830 - 840 - 850 - 860 - 870 - 880 - 890 - 900 - 910
920 - 930 - 940 - 950 - 960 - 970 - 980 - 990 - 1000

Notes: ____

____

____

____

Reg Reps: 10 - 20 - 30 - 40 - 50 - 60 - 70 - 80 - 90 - 100 - 110 - 120 - 130
140 - 150 - 160 - 170 - 180 - 190 - 200 - 210 - 220 - 230 - 240 - 250 - 260
270 - 280 - 290 - 300 - 310 - 320 - 330 - 340 - 350 - 360 - 370 - 380 - 390
400 - 410 - 420 - 430 - 440 - 450 - 460 - 470 - 480 - 490 - 500 - 510 - 520
530 - 540 - 550 - 560 - 570 - 580 - 590 - 600 - 610 - 620 - 630 - 640 - 650
660 - 670 - 680 - 690 - 700 - 710 - 720 - 730 - 740 - 750 - 760 - 770 - 780
790 - 800 - 810 - 820 - 830 - 840 - 850 - 860 - 870 - 880 - 890 - 900 - 910
920 - 930 - 940 - 950 - 960 - 970 - 980 - 990 - 1000

Notes: ____

____

____

____

Reg Reps: 10 - 20 - 30 - 40 - 50 - 60 - 70 - 80 - 90 - 100 - 110 - 120 - 130
140 - 150 - 160 - 170 - 180 - 190 - 200 - 210 - 220 - 230 - 240 - 250 - 260
270 - 280 - 290 - 300 - 310 - 320 - 330 - 340 - 350 - 360 - 370 - 380 - 390
400 - 410 - 420 - 430 - 440 - 450 - 460 - 470 - 480 - 490 - 500 - 510 - 520
530 - 540 - 550 - 560 - 570 - 580 - 590 - 600 - 610 - 620 - 630 - 640 - 650
660 - 670 - 680 - 690 - 700 - 710 - 720 - 730 - 740 - 750 - 760 - 770 - 780
790 - 800 - 810 - 820 - 830 - 840 - 850 - 860 - 870 - 880 - 890 - 900 - 910
920 - 930 - 940 - 950 - 960 - 970 - 980 - 990 - 1000

Notes: ____

____

____

____

Reg Reps: 10 - 20 - 30 - 40 - 50 - 60 - 70 - 80 - 90 - 100 - 110 - 120 - 130
140 - 150 - 160 - 170 - 180 - 190 - 200 - 210 - 220 - 230 - 240 - 250 - 260
270 - 280 - 290 - 300 - 310 - 320 - 330 - 340 - 350 - 360 - 370 - 380 - 390
400 - 410 - 420 - 430 - 440 - 450 - 460 - 470 - 480 - 490 - 500 - 510 - 520
530 - 540 - 550 - 560 - 570 - 580 - 590 - 600 - 610 - 620 - 630 - 640 - 650
660 - 670 - 680 - 690 - 700 - 710 - 720 - 730 - 740 - 750 - 760 - 770 - 780
790 - 800 - 810 - 820 - 830 - 840 - 850 - 860 - 870 - 880 - 890 - 900 - 910
920 - 930 - 940 - 950 - 960 - 970 - 980 - 990 - 1000

Notes: ____

____

____

____

Reg Reps: 10 - 20 - 30 - 40 - 50 - 60 - 70 - 80 - 90 - 100 - 110 - 120 - 130
140 - 150 - 160 - 170 - 180 - 190 - 200 - 210 - 220 - 230 - 240 - 250 - 260
270 - 280 - 290 - 300 - 310 - 320 - 330 - 340 - 350 - 360 - 370 - 380 - 390
400 - 410 - 420 - 430 - 440 - 450 - 460 - 470 - 480 - 490 - 500 - 510 - 520
530 - 540 - 550 - 560 - 570 - 580 - 590 - 600 - 610 - 620 - 630 - 640 - 650
660 - 670 - 680 - 690 - 700 - 710 - 720 - 730 - 740 - 750 - 760 - 770 - 780
790 - 800 - 810 - 820 - 830 - 840 - 850 - 860 - 870 - 880 - 890 - 900 - 910
920 - 930 - 940 - 950 - 960 - 970 - 980 - 990 - 1000

Notes: ____

____

____

______________________________ **End Date:** ____________

Kick: ______________________________ **Start Date:** __________

Kicking Leg: Left - Right - Alt. **Kicking Height: Low - Mid - High**

Notes: ______________________________

____

____

Slow Reps: 10 - 20 - 30 - 40 - 50 - 60 - 70 - 80 - 90 - 100 - 110 - 120 - 130
140 - 150 - 160 - 170 - 180 - 190 - 200 - 210 - 220 - 230 - 240 - 250 - 260
270 - 280 - 290 - 300 - 310 - 320 - 330 - 340 - 350 - 360 - 370 - 380 - 390
400 - 410 - 420 - 430 - 440 - 450 - 460 - 470 - 480 - 490 - 500 - 510 - 520
530 - 540 - 550 - 560 - 570 - 580 - 590 - 600 - 610 - 620 - 630 - 640 - 650
660 - 670 - 680 - 690 - 700 - 710 - 720 - 730 - 740 - 750 - 760 - 770 - 780
790 - 800 - 810 - 820 - 830 - 840 - 850 - 860 - 870 - 880 - 890 - 900 - 910
920 - 930 - 940 - 950 - 960 - 970 - 980 - 990 - 1000

Reg Reps: 10 - 20 - 30 - 40 - 50 - 60 - 70 - 80 - 90 - 100 - 110 - 120 - 130
140 - 150 - 160 - 170 - 180 - 190 - 200 - 210 - 220 - 230 - 240 - 250 - 260
270 - 280 - 290 - 300 - 310 - 320 - 330 - 340 - 350 - 360 - 370 - 380 - 390
400 - 410 - 420 - 430 - 440 - 450 - 460 - 470 - 480 - 490 - 500 - 510 - 520
530 - 540 - 550 - 560 - 570 - 580 - 590 - 600 - 610 - 620 - 630 - 640 - 650
660 - 670 - 680 - 690 - 700 - 710 - 720 - 730 - 740 - 750 - 760 - 770 - 780
790 - 800 - 810 - 820 - 830 - 840 - 850 - 860 - 870 - 880 - 890 - 900 - 910
920 - 930 - 940 - 950 - 960 - 970 - 980 - 990 - 1000

Notes: ______________________________

____

____

____

Reg Reps: 10 - 20 - 30 - 40 - 50 - 60 - 70 - 80 - 90 - 100 - 110 - 120 - 130
140 - 150 - 160 - 170 - 180 - 190 - 200 - 210 - 220 - 230 - 240 - 250 - 260
270 - 280 - 290 - 300 - 310 - 320 - 330 - 340 - 350 - 360 - 370 - 380 - 390
400 - 410 - 420 - 430 - 440 - 450 - 460 - 470 - 480 - 490 - 500 - 510 - 520
530 - 540 - 550 - 560 - 570 - 580 - 590 - 600 - 610 - 620 - 630 - 640 - 650
660 - 670 - 680 - 690 - 700 - 710 - 720 - 730 - 740 - 750 - 760 - 770 - 780
790 - 800 - 810 - 820 - 830 - 840 - 850 - 860 - 870 - 880 - 890 - 900 - 910
920 - 930 - 940 - 950 - 960 - 970 - 980 - 990 - 1000

Notes: ______________________________

____

____

____

Reg Reps: 10 - 20 - 30 - 40 - 50 - 60 - 70 - 80 - 90 - 100 - 110 - 120 - 130
140 - 150 - 160 - 170 - 180 - 190 - 200 - 210 - 220 - 230 - 240 - 250 - 260
270 - 280 - 290 - 300 - 310 - 320 - 330 - 340 - 350 - 360 - 370 - 380 - 390
400 - 410 - 420 - 430 - 440 - 450 - 460 - 470 - 480 - 490 - 500 - 510 - 520
530 - 540 - 550 - 560 - 570 - 580 - 590 - 600 - 610 - 620 - 630 - 640 - 650
660 - 670 - 680 - 690 - 700 - 710 - 720 - 730 - 740 - 750 - 760 - 770 - 780
790 - 800 - 810 - 820 - 830 - 840 - 850 - 860 - 870 - 880 - 890 - 900 - 910
920 - 930 - 940 - 950 - 960 - 970 - 980 - 990 - 1000

Notes: ____

Reg Reps: 10 - 20 - 30 - 40 - 50 - 60 - 70 - 80 - 90 - 100 - 110 - 120 - 130
140 - 150 - 160 - 170 - 180 - 190 - 200 - 210 - 220 - 230 - 240 - 250 - 260
270 - 280 - 290 - 300 - 310 - 320 - 330 - 340 - 350 - 360 - 370 - 380 - 390
400 - 410 - 420 - 430 - 440 - 450 - 460 - 470 - 480 - 490 - 500 - 510 - 520
530 - 540 - 550 - 560 - 570 - 580 - 590 - 600 - 610 - 620 - 630 - 640 - 650
660 - 670 - 680 - 690 - 700 - 710 - 720 - 730 - 740 - 750 - 760 - 770 - 780
790 - 800 - 810 - 820 - 830 - 840 - 850 - 860 - 870 - 880 - 890 - 900 - 910
920 - 930 - 940 - 950 - 960 - 970 - 980 - 990 - 1000

Notes: ____

Reg Reps: 10 - 20 - 30 - 40 - 50 - 60 - 70 - 80 - 90 - 100 - 110 - 120 - 130
140 - 150 - 160 - 170 - 180 - 190 - 200 - 210 - 220 - 230 - 240 - 250 - 260
270 - 280 - 290 - 300 - 310 - 320 - 330 - 340 - 350 - 360 - 370 - 380 - 390
400 - 410 - 420 - 430 - 440 - 450 - 460 - 470 - 480 - 490 - 500 - 510 - 520
530 - 540 - 550 - 560 - 570 - 580 - 590 - 600 - 610 - 620 - 630 - 640 - 650
660 - 670 - 680 - 690 - 700 - 710 - 720 - 730 - 740 - 750 - 760 - 770 - 780
790 - 800 - 810 - 820 - 830 - 840 - 850 - 860 - 870 - 880 - 890 - 900 - 910
920 - 930 - 940 - 950 - 960 - 970 - 980 - 990 - 1000

Notes: ____

End Date: ____________

Kick:_______________________________________ **Start Date:** __________

Kicking Leg: Left - Right - Alt. **Kicking Height: Low - Mid - High**

Notes:____

____

____

Slow Reps: 10 - 20 - 30 - 40 - 50 - 60 - 70 - 80 - 90 - 100 - 110 - 120 - 130
140 - 150 - 160 - 170 - 180 - 190 - 200 - 210 - 220 - 230 - 240 - 250 - 260
270 - 280 - 290 - 300 - 310 - 320 - 330 - 340 - 350 - 360 - 370 - 380 - 390
400 - 410 - 420 - 430 - 440 - 450 - 460 - 470 - 480 - 490 - 500 - 510 - 520
530 - 540 - 550 - 560 - 570 - 580 - 590 - 600 - 610 - 620 - 630 - 640 - 650
660 - 670 - 680 - 690 - 700 - 710 - 720 - 730 - 740 - 750 - 760 - 770 - 780
790 - 800 - 810 - 820 - 830 - 840 - 850 - 860 - 870 - 880 - 890 - 900 - 910
920 - 930 - 940 - 950 - 960 - 970 - 980 - 990 - 1000

Reg Reps: 10 - 20 - 30 - 40 - 50 - 60 - 70 - 80 - 90 - 100 - 110 - 120 - 130
140 - 150 - 160 - 170 - 180 - 190 - 200 - 210 - 220 - 230 - 240 - 250 - 260
270 - 280 - 290 - 300 - 310 - 320 - 330 - 340 - 350 - 360 - 370 - 380 - 390
400 - 410 - 420 - 430 - 440 - 450 - 460 - 470 - 480 - 490 - 500 - 510 - 520
530 - 540 - 550 - 560 - 570 - 580 - 590 - 600 - 610 - 620 - 630 - 640 - 650
660 - 670 - 680 - 690 - 700 - 710 - 720 - 730 - 740 - 750 - 760 - 770 - 780
790 - 800 - 810 - 820 - 830 - 840 - 850 - 860 - 870 - 880 - 890 - 900 - 910
920 - 930 - 940 - 950 - 960 - 970 - 980 - 990 - 1000

Notes: ____

____

____

____

Reg Reps: 10 - 20 - 30 - 40 - 50 - 60 - 70 - 80 - 90 - 100 - 110 - 120 - 130
140 - 150 - 160 - 170 - 180 - 190 - 200 - 210 - 220 - 230 - 240 - 250 - 260
270 - 280 - 290 - 300 - 310 - 320 - 330 - 340 - 350 - 360 - 370 - 380 - 390
400 - 410 - 420 - 430 - 440 - 450 - 460 - 470 - 480 - 490 - 500 - 510 - 520
530 - 540 - 550 - 560 - 570 - 580 - 590 - 600 - 610 - 620 - 630 - 640 - 650
660 - 670 - 680 - 690 - 700 - 710 - 720 - 730 - 740 - 750 - 760 - 770 - 780
790 - 800 - 810 - 820 - 830 - 840 - 850 - 860 - 870 - 880 - 890 - 900 - 910
920 - 930 - 940 - 950 - 960 - 970 - 980 - 990 - 1000

Notes: ____

____

____

____

Reg Reps: 10 - 20 - 30 - 40 - 50 - 60 - 70 - 80 - 90 - 100 - 110 - 120 - 130
140 - 150 - 160 - 170 - 180 - 190 - 200 - 210 - 220 - 230 - 240 - 250 - 260
270 - 280 - 290 - 300 - 310 - 320 - 330 - 340 - 350 - 360 - 370 - 380 - 390
400 - 410 - 420 - 430 - 440 - 450 - 460 - 470 - 480 - 490 - 500 - 510 - 520
530 - 540 - 550 - 560 - 570 - 580 - 590 - 600 - 610 - 620 - 630 - 640 - 650
660 - 670 - 680 - 690 - 700 - 710 - 720 - 730 - 740 - 750 - 760 - 770 - 780
790 - 800 - 810 - 820 - 830 - 840 - 850 - 860 - 870 - 880 - 890 - 900 - 910
920 - 930 - 940 - 950 - 960 - 970 - 980 - 990 - 1000

Notes: ____

____

____

____

Reg Reps: 10 - 20 - 30 - 40 - 50 - 60 - 70 - 80 - 90 - 100 - 110 - 120 - 130
140 - 150 - 160 - 170 - 180 - 190 - 200 - 210 - 220 - 230 - 240 - 250 - 260
270 - 280 - 290 - 300 - 310 - 320 - 330 - 340 - 350 - 360 - 370 - 380 - 390
400 - 410 - 420 - 430 - 440 - 450 - 460 - 470 - 480 - 490 - 500 - 510 - 520
530 - 540 - 550 - 560 - 570 - 580 - 590 - 600 - 610 - 620 - 630 - 640 - 650
660 - 670 - 680 - 690 - 700 - 710 - 720 - 730 - 740 - 750 - 760 - 770 - 780
790 - 800 - 810 - 820 - 830 - 840 - 850 - 860 - 870 - 880 - 890 - 900 - 910
920 - 930 - 940 - 950 - 960 - 970 - 980 - 990 - 1000

Notes: ____

____

____

____

Reg Reps: 10 - 20 - 30 - 40 - 50 - 60 - 70 - 80 - 90 - 100 - 110 - 120 - 130
140 - 150 - 160 - 170 - 180 - 190 - 200 - 210 - 220 - 230 - 240 - 250 - 260
270 - 280 - 290 - 300 - 310 - 320 - 330 - 340 - 350 - 360 - 370 - 380 - 390
400 - 410 - 420 - 430 - 440 - 450 - 460 - 470 - 480 - 490 - 500 - 510 - 520
530 - 540 - 550 - 560 - 570 - 580 - 590 - 600 - 610 - 620 - 630 - 640 - 650
660 - 670 - 680 - 690 - 700 - 710 - 720 - 730 - 740 - 750 - 760 - 770 - 780
790 - 800 - 810 - 820 - 830 - 840 - 850 - 860 - 870 - 880 - 890 - 900 - 910
920 - 930 - 940 - 950 - 960 - 970 - 980 - 990 - 1000

Notes: ____

____

____

____ **End Date:** __________

Kick:______________________________ **Start Date:** __________

Kicking Leg: Left - Right - Alt. **Kicking Height: Low - Mid - High**

Notes:____

____

____

Slow Reps: 10 - 20 - 30 - 40 - 50 - 60 - 70 - 80 - 90 - 100 - 110 - 120 - 130
140 - 150 - 160 - 170 - 180 - 190 - 200 - 210 - 220 - 230 - 240 - 250 - 260
270 - 280 - 290 - 300 - 310 - 320 - 330 - 340 - 350 - 360 - 370 - 380 - 390
400 - 410 - 420 - 430 - 440 - 450 - 460 - 470 - 480 - 490 - 500 - 510 - 520
530 - 540 - 550 - 560 - 570 - 580 - 590 - 600 - 610 - 620 - 630 - 640 - 650
660 - 670 - 680 - 690 - 700 - 710 - 720 - 730 - 740 - 750 - 760 - 770 - 780
790 - 800 - 810 - 820 - 830 - 840 - 850 - 860 - 870 - 880 - 890 - 900 - 910
920 - 930 - 940 - 950 - 960 - 970 - 980 - 990 - 1000

Reg Reps: 10 - 20 - 30 - 40 - 50 - 60 - 70 - 80 - 90 - 100 - 110 - 120 - 130
140 - 150 - 160 - 170 - 180 - 190 - 200 - 210 - 220 - 230 - 240 - 250 - 260
270 - 280 - 290 - 300 - 310 - 320 - 330 - 340 - 350 - 360 - 370 - 380 - 390
400 - 410 - 420 - 430 - 440 - 450 - 460 - 470 - 480 - 490 - 500 - 510 - 520
530 - 540 - 550 - 560 - 570 - 580 - 590 - 600 - 610 - 620 - 630 - 640 - 650
660 - 670 - 680 - 690 - 700 - 710 - 720 - 730 - 740 - 750 - 760 - 770 - 780
790 - 800 - 810 - 820 - 830 - 840 - 850 - 860 - 870 - 880 - 890 - 900 - 910
920 - 930 - 940 - 950 - 960 - 970 - 980 - 990 - 1000

Notes: ____

____

____

____

Reg Reps: 10 - 20 - 30 - 40 - 50 - 60 - 70 - 80 - 90 - 100 - 110 - 120 - 130
140 - 150 - 160 - 170 - 180 - 190 - 200 - 210 - 220 - 230 - 240 - 250 - 260
270 - 280 - 290 - 300 - 310 - 320 - 330 - 340 - 350 - 360 - 370 - 380 - 390
400 - 410 - 420 - 430 - 440 - 450 - 460 - 470 - 480 - 490 - 500 - 510 - 520
530 - 540 - 550 - 560 - 570 - 580 - 590 - 600 - 610 - 620 - 630 - 640 - 650
660 - 670 - 680 - 690 - 700 - 710 - 720 - 730 - 740 - 750 - 760 - 770 - 780
790 - 800 - 810 - 820 - 830 - 840 - 850 - 860 - 870 - 880 - 890 - 900 - 910
920 - 930 - 940 - 950 - 960 - 970 - 980 - 990 - 1000

Notes: ____

____

____

____

Reg Reps: 10 - 20 - 30 - 40 - 50 - 60 - 70 - 80 - 90 - 100 - 110 - 120 - 130 140 - 150 - 160 - 170 - 180 - 190 - 200 - 210 - 220 - 230 - 240 - 250 - 260 270 - 280 - 290 - 300 - 310 - 320 - 330 - 340 - 350 - 360 - 370 - 380 - 390 400 - 410 - 420 - 430 - 440 - 450 - 460 - 470 - 480 - 490 - 500 - 510 - 520 530 - 540 - 550 - 560 - 570 - 580 - 590 - 600 - 610 - 620 - 630 - 640 - 650 660 - 670 - 680 - 690 - 700 - 710 - 720 - 730 - 740 - 750 - 760 - 770 - 780 790 - 800 - 810 - 820 - 830 - 840 - 850 - 860 - 870 - 880 - 890 - 900 - 910 920 - 930 - 940 - 950 - 960 - 970 - 980 - 990 - 1000

Notes: ____

____

____

____

Reg Reps: 10 - 20 - 30 - 40 - 50 - 60 - 70 - 80 - 90 - 100 - 110 - 120 - 130 140 - 150 - 160 - 170 - 180 - 190 - 200 - 210 - 220 - 230 - 240 - 250 - 260 270 - 280 - 290 - 300 - 310 - 320 - 330 - 340 - 350 - 360 - 370 - 380 - 390 400 - 410 - 420 - 430 - 440 - 450 - 460 - 470 - 480 - 490 - 500 - 510 - 520 530 - 540 - 550 - 560 - 570 - 580 - 590 - 600 - 610 - 620 - 630 - 640 - 650 660 - 670 - 680 - 690 - 700 - 710 - 720 - 730 - 740 - 750 - 760 - 770 - 780 790 - 800 - 810 - 820 - 830 - 840 - 850 - 860 - 870 - 880 - 890 - 900 - 910 920 - 930 - 940 - 950 - 960 - 970 - 980 - 990 - 1000

Notes: ____

____

____

____

Reg Reps: 10 - 20 - 30 - 40 - 50 - 60 - 70 - 80 - 90 - 100 - 110 - 120 - 130 140 - 150 - 160 - 170 - 180 - 190 - 200 - 210 - 220 - 230 - 240 - 250 - 260 270 - 280 - 290 - 300 - 310 - 320 - 330 - 340 - 350 - 360 - 370 - 380 - 390 400 - 410 - 420 - 430 - 440 - 450 - 460 - 470 - 480 - 490 - 500 - 510 - 520 530 - 540 - 550 - 560 - 570 - 580 - 590 - 600 - 610 - 620 - 630 - 640 - 650 660 - 670 - 680 - 690 - 700 - 710 - 720 - 730 - 740 - 750 - 760 - 770 - 780 790 - 800 - 810 - 820 - 830 - 840 - 850 - 860 - 870 - 880 - 890 - 900 - 910 920 - 930 - 940 - 950 - 960 - 970 - 980 - 990 - 1000

Notes: ____

____

____

______________________________ **End Date:** ____________

Kick: ______________________________ **Start Date:** __________

Kicking Leg: Left - Right - Alt. **Kicking Height: Low - Mid - High**

Notes: ______________________________

____

____

Slow Reps: 10 - 20 - 30 - 40 - 50 - 60 - 70 - 80 - 90 - 100 - 110 - 120 - 130
140 - 150 - 160 - 170 - 180 - 190 - 200 - 210 - 220 - 230 - 240 - 250 - 260
270 - 280 - 290 - 300 - 310 - 320 - 330 - 340 - 350 - 360 - 370 - 380 - 390
400 - 410 - 420 - 430 - 440 - 450 - 460 - 470 - 480 - 490 - 500 - 510 - 520
530 - 540 - 550 - 560 - 570 - 580 - 590 - 600 - 610 - 620 - 630 - 640 - 650
660 - 670 - 680 - 690 - 700 - 710 - 720 - 730 - 740 - 750 - 760 - 770 - 780
790 - 800 - 810 - 820 - 830 - 840 - 850 - 860 - 870 - 880 - 890 - 900 - 910
920 - 930 - 940 - 950 - 960 - 970 - 980 - 990 - 1000

Reg Reps: 10 - 20 - 30 - 40 - 50 - 60 - 70 - 80 - 90 - 100 - 110 - 120 - 130
140 - 150 - 160 - 170 - 180 - 190 - 200 - 210 - 220 - 230 - 240 - 250 - 260
270 - 280 - 290 - 300 - 310 - 320 - 330 - 340 - 350 - 360 - 370 - 380 - 390
400 - 410 - 420 - 430 - 440 - 450 - 460 - 470 - 480 - 490 - 500 - 510 - 520
530 - 540 - 550 - 560 - 570 - 580 - 590 - 600 - 610 - 620 - 630 - 640 - 650
660 - 670 - 680 - 690 - 700 - 710 - 720 - 730 - 740 - 750 - 760 - 770 - 780
790 - 800 - 810 - 820 - 830 - 840 - 850 - 860 - 870 - 880 - 890 - 900 - 910
920 - 930 - 940 - 950 - 960 - 970 - 980 - 990 - 1000

Notes: ______________________________

____

____

____

Reg Reps: 10 - 20 - 30 - 40 - 50 - 60 - 70 - 80 - 90 - 100 - 110 - 120 - 130
140 - 150 - 160 - 170 - 180 - 190 - 200 - 210 - 220 - 230 - 240 - 250 - 260
270 - 280 - 290 - 300 - 310 - 320 - 330 - 340 - 350 - 360 - 370 - 380 - 390
400 - 410 - 420 - 430 - 440 - 450 - 460 - 470 - 480 - 490 - 500 - 510 - 520
530 - 540 - 550 - 560 - 570 - 580 - 590 - 600 - 610 - 620 - 630 - 640 - 650
660 - 670 - 680 - 690 - 700 - 710 - 720 - 730 - 740 - 750 - 760 - 770 - 780
790 - 800 - 810 - 820 - 830 - 840 - 850 - 860 - 870 - 880 - 890 - 900 - 910
920 - 930 - 940 - 950 - 960 - 970 - 980 - 990 - 1000

Notes: ______________________________

____

____

____

Reg Reps: 10 - 20 - 30 - 40 - 50 - 60 - 70 - 80 - 90 - 100 - 110 - 120 - 130
140 - 150 - 160 - 170 - 180 - 190 - 200 - 210 - 220 - 230 - 240 - 250 - 260
270 - 280 - 290 - 300 - 310 - 320 - 330 - 340 - 350 - 360 - 370 - 380 - 390
400 - 410 - 420 - 430 - 440 - 450 - 460 - 470 - 480 - 490 - 500 - 510 - 520
530 - 540 - 550 - 560 - 570 - 580 - 590 - 600 - 610 - 620 - 630 - 640 - 650
660 - 670 - 680 - 690 - 700 - 710 - 720 - 730 - 740 - 750 - 760 - 770 - 780
790 - 800 - 810 - 820 - 830 - 840 - 850 - 860 - 870 - 880 - 890 - 900 - 910
920 - 930 - 940 - 950 - 960 - 970 - 980 - 990 - 1000

Notes: ____

____

____

____

Reg Reps: 10 - 20 - 30 - 40 - 50 - 60 - 70 - 80 - 90 - 100 - 110 - 120 - 130
140 - 150 - 160 - 170 - 180 - 190 - 200 - 210 - 220 - 230 - 240 - 250 - 260
270 - 280 - 290 - 300 - 310 - 320 - 330 - 340 - 350 - 360 - 370 - 380 - 390
400 - 410 - 420 - 430 - 440 - 450 - 460 - 470 - 480 - 490 - 500 - 510 - 520
530 - 540 - 550 - 560 - 570 - 580 - 590 - 600 - 610 - 620 - 630 - 640 - 650
660 - 670 - 680 - 690 - 700 - 710 - 720 - 730 - 740 - 750 - 760 - 770 - 780
790 - 800 - 810 - 820 - 830 - 840 - 850 - 860 - 870 - 880 - 890 - 900 - 910
920 - 930 - 940 - 950 - 960 - 970 - 980 - 990 - 1000

Notes: ____

____

____

____

Reg Reps: 10 - 20 - 30 - 40 - 50 - 60 - 70 - 80 - 90 - 100 - 110 - 120 - 130
140 - 150 - 160 - 170 - 180 - 190 - 200 - 210 - 220 - 230 - 240 - 250 - 260
270 - 280 - 290 - 300 - 310 - 320 - 330 - 340 - 350 - 360 - 370 - 380 - 390
400 - 410 - 420 - 430 - 440 - 450 - 460 - 470 - 480 - 490 - 500 - 510 - 520
530 - 540 - 550 - 560 - 570 - 580 - 590 - 600 - 610 - 620 - 630 - 640 - 650
660 - 670 - 680 - 690 - 700 - 710 - 720 - 730 - 740 - 750 - 760 - 770 - 780
790 - 800 - 810 - 820 - 830 - 840 - 850 - 860 - 870 - 880 - 890 - 900 - 910
920 - 930 - 940 - 950 - 960 - 970 - 980 - 990 - 1000

Notes: ____

____

____

______________________________________ **End Date:** ____________

Kick:______________________________ **Start Date:** __________

Kicking Leg: Left - Right - Alt. **Kicking Height: Low - Mid - High**

Notes:____

Slow Reps: 10 - 20 - 30 - 40 - 50 - 60 - 70 - 80 - 90 - 100 - 110 - 120 - 130
140 - 150 - 160 - 170 - 180 - 190 - 200 - 210 - 220 - 230 - 240 - 250 - 260
270 - 280 - 290 - 300 - 310 - 320 - 330 - 340 - 350 - 360 - 370 - 380 - 390
400 - 410 - 420 - 430 - 440 - 450 - 460 - 470 - 480 - 490 - 500 - 510 - 520
530 - 540 - 550 - 560 - 570 - 580 - 590 - 600 - 610 - 620 - 630 - 640 - 650
660 - 670 - 680 - 690 - 700 - 710 - 720 - 730 - 740 - 750 - 760 - 770 - 780
790 - 800 - 810 - 820 - 830 - 840 - 850 - 860 - 870 - 880 - 890 - 900 - 910
920 - 930 - 940 - 950 - 960 - 970 - 980 - 990 - 1000

Reg Reps: 10 - 20 - 30 - 40 - 50 - 60 - 70 - 80 - 90 - 100 - 110 - 120 - 130
140 - 150 - 160 - 170 - 180 - 190 - 200 - 210 - 220 - 230 - 240 - 250 - 260
270 - 280 - 290 - 300 - 310 - 320 - 330 - 340 - 350 - 360 - 370 - 380 - 390
400 - 410 - 420 - 430 - 440 - 450 - 460 - 470 - 480 - 490 - 500 - 510 - 520
530 - 540 - 550 - 560 - 570 - 580 - 590 - 600 - 610 - 620 - 630 - 640 - 650
660 - 670 - 680 - 690 - 700 - 710 - 720 - 730 - 740 - 750 - 760 - 770 - 780
790 - 800 - 810 - 820 - 830 - 840 - 850 - 860 - 870 - 880 - 890 - 900 - 910
920 - 930 - 940 - 950 - 960 - 970 - 980 - 990 - 1000

Notes: ____

Reg Reps: 10 - 20 - 30 - 40 - 50 - 60 - 70 - 80 - 90 - 100 - 110 - 120 - 130
140 - 150 - 160 - 170 - 180 - 190 - 200 - 210 - 220 - 230 - 240 - 250 - 260
270 - 280 - 290 - 300 - 310 - 320 - 330 - 340 - 350 - 360 - 370 - 380 - 390
400 - 410 - 420 - 430 - 440 - 450 - 460 - 470 - 480 - 490 - 500 - 510 - 520
530 - 540 - 550 - 560 - 570 - 580 - 590 - 600 - 610 - 620 - 630 - 640 - 650
660 - 670 - 680 - 690 - 700 - 710 - 720 - 730 - 740 - 750 - 760 - 770 - 780
790 - 800 - 810 - 820 - 830 - 840 - 850 - 860 - 870 - 880 - 890 - 900 - 910
920 - 930 - 940 - 950 - 960 - 970 - 980 - 990 - 1000

Notes: ____

Reg Reps: 10 - 20 - 30 - 40 - 50 - 60 - 70 - 80 - 90 - 100 - 110 - 120 - 130 - 140 - 150 - 160 - 170 - 180 - 190 - 200 - 210 - 220 - 230 - 240 - 250 - 260 - 270 - 280 - 290 - 300 - 310 - 320 - 330 - 340 - 350 - 360 - 370 - 380 - 390 - 400 - 410 - 420 - 430 - 440 - 450 - 460 - 470 - 480 - 490 - 500 - 510 - 520 - 530 - 540 - 550 - 560 - 570 - 580 - 590 - 600 - 610 - 620 - 630 - 640 - 650 - 660 - 670 - 680 - 690 - 700 - 710 - 720 - 730 - 740 - 750 - 760 - 770 - 780 - 790 - 800 - 810 - 820 - 830 - 840 - 850 - 860 - 870 - 880 - 890 - 900 - 910 - 920 - 930 - 940 - 950 - 960 - 970 - 980 - 990 - 1000

Notes: ____

____

____

____

Reg Reps: 10 - 20 - 30 - 40 - 50 - 60 - 70 - 80 - 90 - 100 - 110 - 120 - 130 - 140 - 150 - 160 - 170 - 180 - 190 - 200 - 210 - 220 - 230 - 240 - 250 - 260 - 270 - 280 - 290 - 300 - 310 - 320 - 330 - 340 - 350 - 360 - 370 - 380 - 390 - 400 - 410 - 420 - 430 - 440 - 450 - 460 - 470 - 480 - 490 - 500 - 510 - 520 - 530 - 540 - 550 - 560 - 570 - 580 - 590 - 600 - 610 - 620 - 630 - 640 - 650 - 660 - 670 - 680 - 690 - 700 - 710 - 720 - 730 - 740 - 750 - 760 - 770 - 780 - 790 - 800 - 810 - 820 - 830 - 840 - 850 - 860 - 870 - 880 - 890 - 900 - 910 - 920 - 930 - 940 - 950 - 960 - 970 - 980 - 990 - 1000

Notes: ____

____

____

____

Reg Reps: 10 - 20 - 30 - 40 - 50 - 60 - 70 - 80 - 90 - 100 - 110 - 120 - 130 - 140 - 150 - 160 - 170 - 180 - 190 - 200 - 210 - 220 - 230 - 240 - 250 - 260 - 270 - 280 - 290 - 300 - 310 - 320 - 330 - 340 - 350 - 360 - 370 - 380 - 390 - 400 - 410 - 420 - 430 - 440 - 450 - 460 - 470 - 480 - 490 - 500 - 510 - 520 - 530 - 540 - 550 - 560 - 570 - 580 - 590 - 600 - 610 - 620 - 630 - 640 - 650 - 660 - 670 - 680 - 690 - 700 - 710 - 720 - 730 - 740 - 750 - 760 - 770 - 780 - 790 - 800 - 810 - 820 - 830 - 840 - 850 - 860 - 870 - 880 - 890 - 900 - 910 - 920 - 930 - 940 - 950 - 960 - 970 - 980 - 990 - 1000

Notes: ____

____

____

______________________________ **End Date:** ____________

Kick:______________________________ **Start Date:** _________
Kicking Leg: Left - Right - Alt. **Kicking Height: Low - Mid - High**
Notes:____

Slow Reps: 10 - 20 - 30 - 40 - 50 - 60 - 70 - 80 - 90 - 100 - 110 - 120 - 130
140 - 150 - 160 - 170 - 180 - 190 - 200 - 210 - 220 - 230 - 240 - 250 - 260
270 - 280 - 290 - 300 - 310 - 320 - 330 - 340 - 350 - 360 - 370 - 380 - 390
400 - 410 - 420 - 430 - 440 - 450 - 460 - 470 - 480 - 490 - 500 - 510 - 520
530 - 540 - 550 - 560 - 570 - 580 - 590 - 600 - 610 - 620 - 630 - 640 - 650
660 - 670 - 680 - 690 - 700 - 710 - 720 - 730 - 740 - 750 - 760 - 770 - 780
790 - 800 - 810 - 820 - 830 - 840 - 850 - 860 - 870 - 880 - 890 - 900 - 910
920 - 930 - 940 - 950 - 960 - 970 - 980 - 990 - 1000

Reg Reps: 10 - 20 - 30 - 40 - 50 - 60 - 70 - 80 - 90 - 100 - 110 - 120 - 130
140 - 150 - 160 - 170 - 180 - 190 - 200 - 210 - 220 - 230 - 240 - 250 - 260
270 - 280 - 290 - 300 - 310 - 320 - 330 - 340 - 350 - 360 - 370 - 380 - 390
400 - 410 - 420 - 430 - 440 - 450 - 460 - 470 - 480 - 490 - 500 - 510 - 520
530 - 540 - 550 - 560 - 570 - 580 - 590 - 600 - 610 - 620 - 630 - 640 - 650
660 - 670 - 680 - 690 - 700 - 710 - 720 - 730 - 740 - 750 - 760 - 770 - 780
790 - 800 - 810 - 820 - 830 - 840 - 850 - 860 - 870 - 880 - 890 - 900 - 910
920 - 930 - 940 - 950 - 960 - 970 - 980 - 990 - 1000
Notes: ____

Reg Reps: 10 - 20 - 30 - 40 - 50 - 60 - 70 - 80 - 90 - 100 - 110 - 120 - 130
140 - 150 - 160 - 170 - 180 - 190 - 200 - 210 - 220 - 230 - 240 - 250 - 260
270 - 280 - 290 - 300 - 310 - 320 - 330 - 340 - 350 - 360 - 370 - 380 - 390
400 - 410 - 420 - 430 - 440 - 450 - 460 - 470 - 480 - 490 - 500 - 510 - 520
530 - 540 - 550 - 560 - 570 - 580 - 590 - 600 - 610 - 620 - 630 - 640 - 650
660 - 670 - 680 - 690 - 700 - 710 - 720 - 730 - 740 - 750 - 760 - 770 - 780
790 - 800 - 810 - 820 - 830 - 840 - 850 - 860 - 870 - 880 - 890 - 900 - 910
920 - 930 - 940 - 950 - 960 - 970 - 980 - 990 - 1000
Notes: ____

Reg Reps: 10 - 20 - 30 - 40 - 50 - 60 - 70 - 80 - 90 - 100 - 110 - 120 - 130
140 - 150 - 160 - 170 - 180 - 190 - 200 - 210 - 220 - 230 - 240 - 250 - 260
270 - 280 - 290 - 300 - 310 - 320 - 330 - 340 - 350 - 360 - 370 - 380 - 390
400 - 410 - 420 - 430 - 440 - 450 - 460 - 470 - 480 - 490 - 500 - 510 - 520
530 - 540 - 550 - 560 - 570 - 580 - 590 - 600 - 610 - 620 - 630 - 640 - 650
660 - 670 - 680 - 690 - 700 - 710 - 720 - 730 - 740 - 750 - 760 - 770 - 780
790 - 800 - 810 - 820 - 830 - 840 - 850 - 860 - 870 - 880 - 890 - 900 - 910
920 - 930 - 940 - 950 - 960 - 970 - 980 - 990 - 1000

Notes: ____

____

____

____

Reg Reps: 10 - 20 - 30 - 40 - 50 - 60 - 70 - 80 - 90 - 100 - 110 - 120 - 130
140 - 150 - 160 - 170 - 180 - 190 - 200 - 210 - 220 - 230 - 240 - 250 - 260
270 - 280 - 290 - 300 - 310 - 320 - 330 - 340 - 350 - 360 - 370 - 380 - 390
400 - 410 - 420 - 430 - 440 - 450 - 460 - 470 - 480 - 490 - 500 - 510 - 520
530 - 540 - 550 - 560 - 570 - 580 - 590 - 600 - 610 - 620 - 630 - 640 - 650
660 - 670 - 680 - 690 - 700 - 710 - 720 - 730 - 740 - 750 - 760 - 770 - 780
790 - 800 - 810 - 820 - 830 - 840 - 850 - 860 - 870 - 880 - 890 - 900 - 910
920 - 930 - 940 - 950 - 960 - 970 - 980 - 990 - 1000

Notes: ____

____

____

____

Reg Reps: 10 - 20 - 30 - 40 - 50 - 60 - 70 - 80 - 90 - 100 - 110 - 120 - 130
140 - 150 - 160 - 170 - 180 - 190 - 200 - 210 - 220 - 230 - 240 - 250 - 260
270 - 280 - 290 - 300 - 310 - 320 - 330 - 340 - 350 - 360 - 370 - 380 - 390
400 - 410 - 420 - 430 - 440 - 450 - 460 - 470 - 480 - 490 - 500 - 510 - 520
530 - 540 - 550 - 560 - 570 - 580 - 590 - 600 - 610 - 620 - 630 - 640 - 650
660 - 670 - 680 - 690 - 700 - 710 - 720 - 730 - 740 - 750 - 760 - 770 - 780
790 - 800 - 810 - 820 - 830 - 840 - 850 - 860 - 870 - 880 - 890 - 900 - 910
920 - 930 - 940 - 950 - 960 - 970 - 980 - 990 - 1000

Notes: ____

____

____

______________________________ **End Date:** ______________

Kick:__________________________________ **Start Date:** _________

Kicking Leg: Left - Right - Alt. **Kicking Height: Low - Mid - High**

Notes:____

Slow Reps: 10 - 20 - 30 - 40 - 50 - 60 - 70 - 80 - 90 - 100 - 110 - 120 - 130
140 - 150 - 160 - 170 - 180 - 190 - 200 - 210 - 220 - 230 - 240 - 250 - 260
270 - 280 - 290 - 300 - 310 - 320 - 330 - 340 - 350 - 360 - 370 - 380 - 390
400 - 410 - 420 - 430 - 440 - 450 - 460 - 470 - 480 - 490 - 500 - 510 - 520
530 - 540 - 550 - 560 - 570 - 580 - 590 - 600 - 610 - 620 - 630 - 640 - 650
660 - 670 - 680 - 690 - 700 - 710 - 720 - 730 - 740 - 750 - 760 - 770 - 780
790 - 800 - 810 - 820 - 830 - 840 - 850 - 860 - 870 - 880 - 890 - 900 - 910
920 - 930 - 940 - 950 - 960 - 970 - 980 - 990 - 1000

Reg Reps: 10 - 20 - 30 - 40 - 50 - 60 - 70 - 80 - 90 - 100 - 110 - 120 - 130
140 - 150 - 160 - 170 - 180 - 190 - 200 - 210 - 220 - 230 - 240 - 250 - 260
270 - 280 - 290 - 300 - 310 - 320 - 330 - 340 - 350 - 360 - 370 - 380 - 390
400 - 410 - 420 - 430 - 440 - 450 - 460 - 470 - 480 - 490 - 500 - 510 - 520
530 - 540 - 550 - 560 - 570 - 580 - 590 - 600 - 610 - 620 - 630 - 640 - 650
660 - 670 - 680 - 690 - 700 - 710 - 720 - 730 - 740 - 750 - 760 - 770 - 780
790 - 800 - 810 - 820 - 830 - 840 - 850 - 860 - 870 - 880 - 890 - 900 - 910
920 - 930 - 940 - 950 - 960 - 970 - 980 - 990 - 1000

Notes: ____

Reg Reps: 10 - 20 - 30 - 40 - 50 - 60 - 70 - 80 - 90 - 100 - 110 - 120 - 130
140 - 150 - 160 - 170 - 180 - 190 - 200 - 210 - 220 - 230 - 240 - 250 - 260
270 - 280 - 290 - 300 - 310 - 320 - 330 - 340 - 350 - 360 - 370 - 380 - 390
400 - 410 - 420 - 430 - 440 - 450 - 460 - 470 - 480 - 490 - 500 - 510 - 520
530 - 540 - 550 - 560 - 570 - 580 - 590 - 600 - 610 - 620 - 630 - 640 - 650
660 - 670 - 680 - 690 - 700 - 710 - 720 - 730 - 740 - 750 - 760 - 770 - 780
790 - 800 - 810 - 820 - 830 - 840 - 850 - 860 - 870 - 880 - 890 - 900 - 910
920 - 930 - 940 - 950 - 960 - 970 - 980 - 990 - 1000

Notes: ____

Reg Reps: 10 - 20 - 30 - 40 - 50 - 60 - 70 - 80 - 90 - 100 - 110 - 120 - 130
140 - 150 - 160 - 170 - 180 - 190 - 200 - 210 - 220 - 230 - 240 - 250 - 260
270 - 280 - 290 - 300 - 310 - 320 - 330 - 340 - 350 - 360 - 370 - 380 - 390
400 - 410 - 420 - 430 - 440 - 450 - 460 - 470 - 480 - 490 - 500 - 510 - 520
530 - 540 - 550 - 560 - 570 - 580 - 590 - 600 - 610 - 620 - 630 - 640 - 650
660 - 670 - 680 - 690 - 700 - 710 - 720 - 730 - 740 - 750 - 760 - 770 - 780
790 - 800 - 810 - 820 - 830 - 840 - 850 - 860 - 870 - 880 - 890 - 900 - 910
920 - 930 - 940 - 950 - 960 - 970 - 980 - 990 - 1000

Notes: ____

Reg Reps: 10 - 20 - 30 - 40 - 50 - 60 - 70 - 80 - 90 - 100 - 110 - 120 - 130
140 - 150 - 160 - 170 - 180 - 190 - 200 - 210 - 220 - 230 - 240 - 250 - 260
270 - 280 - 290 - 300 - 310 - 320 - 330 - 340 - 350 - 360 - 370 - 380 - 390
400 - 410 - 420 - 430 - 440 - 450 - 460 - 470 - 480 - 490 - 500 - 510 - 520
530 - 540 - 550 - 560 - 570 - 580 - 590 - 600 - 610 - 620 - 630 - 640 - 650
660 - 670 - 680 - 690 - 700 - 710 - 720 - 730 - 740 - 750 - 760 - 770 - 780
790 - 800 - 810 - 820 - 830 - 840 - 850 - 860 - 870 - 880 - 890 - 900 - 910
920 - 930 - 940 - 950 - 960 - 970 - 980 - 990 - 1000

Notes: ____

Reg Reps: 10 - 20 - 30 - 40 - 50 - 60 - 70 - 80 - 90 - 100 - 110 - 120 - 130
140 - 150 - 160 - 170 - 180 - 190 - 200 - 210 - 220 - 230 - 240 - 250 - 260
270 - 280 - 290 - 300 - 310 - 320 - 330 - 340 - 350 - 360 - 370 - 380 - 390
400 - 410 - 420 - 430 - 440 - 450 - 460 - 470 - 480 - 490 - 500 - 510 - 520
530 - 540 - 550 - 560 - 570 - 580 - 590 - 600 - 610 - 620 - 630 - 640 - 650
660 - 670 - 680 - 690 - 700 - 710 - 720 - 730 - 740 - 750 - 760 - 770 - 780
790 - 800 - 810 - 820 - 830 - 840 - 850 - 860 - 870 - 880 - 890 - 900 - 910
920 - 930 - 940 - 950 - 960 - 970 - 980 - 990 - 1000

Notes: ____

End Date: ____________

Kick:____________________________________ **Start Date:** __________
Kicking Leg: Left - Right - Alt. **Kicking Height: Low - Mid - High**
Notes:____

____

____

Slow Reps: 10 - 20 - 30 - 40 - 50 - 60 - 70 - 80 - 90 - 100 - 110 - 120 - 130
140 - 150 - 160 - 170 - 180 - 190 - 200 - 210 - 220 - 230 - 240 - 250 - 260
270 - 280 - 290 - 300 - 310 - 320 - 330 - 340 - 350 - 360 - 370 - 380 - 390
400 - 410 - 420 - 430 - 440 - 450 - 460 - 470 - 480 - 490 - 500 - 510 - 520
530 - 540 - 550 - 560 - 570 - 580 - 590 - 600 - 610 - 620 - 630 - 640 - 650
660 - 670 - 680 - 690 - 700 - 710 - 720 - 730 - 740 - 750 - 760 - 770 - 780
790 - 800 - 810 - 820 - 830 - 840 - 850 - 860 - 870 - 880 - 890 - 900 - 910
920 - 930 - 940 - 950 - 960 - 970 - 980 - 990 - 1000

Reg Reps: 10 - 20 - 30 - 40 - 50 - 60 - 70 - 80 - 90 - 100 - 110 - 120 - 130
140 - 150 - 160 - 170 - 180 - 190 - 200 - 210 - 220 - 230 - 240 - 250 - 260
270 - 280 - 290 - 300 - 310 - 320 - 330 - 340 - 350 - 360 - 370 - 380 - 390
400 - 410 - 420 - 430 - 440 - 450 - 460 - 470 - 480 - 490 - 500 - 510 - 520
530 - 540 - 550 - 560 - 570 - 580 - 590 - 600 - 610 - 620 - 630 - 640 - 650
660 - 670 - 680 - 690 - 700 - 710 - 720 - 730 - 740 - 750 - 760 - 770 - 780
790 - 800 - 810 - 820 - 830 - 840 - 850 - 860 - 870 - 880 - 890 - 900 - 910
920 - 930 - 940 - 950 - 960 - 970 - 980 - 990 - 1000
Notes: ____

____

____

____

Reg Reps: 10 - 20 - 30 - 40 - 50 - 60 - 70 - 80 - 90 - 100 - 110 - 120 - 130
140 - 150 - 160 - 170 - 180 - 190 - 200 - 210 - 220 - 230 - 240 - 250 - 260
270 - 280 - 290 - 300 - 310 - 320 - 330 - 340 - 350 - 360 - 370 - 380 - 390
400 - 410 - 420 - 430 - 440 - 450 - 460 - 470 - 480 - 490 - 500 - 510 - 520
530 - 540 - 550 - 560 - 570 - 580 - 590 - 600 - 610 - 620 - 630 - 640 - 650
660 - 670 - 680 - 690 - 700 - 710 - 720 - 730 - 740 - 750 - 760 - 770 - 780
790 - 800 - 810 - 820 - 830 - 840 - 850 - 860 - 870 - 880 - 890 - 900 - 910
920 - 930 - 940 - 950 - 960 - 970 - 980 - 990 - 1000
Notes: ____

____

____

____

Reg Reps: 10 - 20 - 30 - 40 - 50 - 60 - 70 - 80 - 90 - 100 - 110 - 120 - 130 - 140 - 150 - 160 - 170 - 180 - 190 - 200 - 210 - 220 - 230 - 240 - 250 - 260 - 270 - 280 - 290 - 300 - 310 - 320 - 330 - 340 - 350 - 360 - 370 - 380 - 390 - 400 - 410 - 420 - 430 - 440 - 450 - 460 - 470 - 480 - 490 - 500 - 510 - 520 - 530 - 540 - 550 - 560 - 570 - 580 - 590 - 600 - 610 - 620 - 630 - 640 - 650 - 660 - 670 - 680 - 690 - 700 - 710 - 720 - 730 - 740 - 750 - 760 - 770 - 780 - 790 - 800 - 810 - 820 - 830 - 840 - 850 - 860 - 870 - 880 - 890 - 900 - 910 - 920 - 930 - 940 - 950 - 960 - 970 - 980 - 990 - 1000

Notes: ____

____

____

____

Reg Reps: 10 - 20 - 30 - 40 - 50 - 60 - 70 - 80 - 90 - 100 - 110 - 120 - 130 - 140 - 150 - 160 - 170 - 180 - 190 - 200 - 210 - 220 - 230 - 240 - 250 - 260 - 270 - 280 - 290 - 300 - 310 - 320 - 330 - 340 - 350 - 360 - 370 - 380 - 390 - 400 - 410 - 420 - 430 - 440 - 450 - 460 - 470 - 480 - 490 - 500 - 510 - 520 - 530 - 540 - 550 - 560 - 570 - 580 - 590 - 600 - 610 - 620 - 630 - 640 - 650 - 660 - 670 - 680 - 690 - 700 - 710 - 720 - 730 - 740 - 750 - 760 - 770 - 780 - 790 - 800 - 810 - 820 - 830 - 840 - 850 - 860 - 870 - 880 - 890 - 900 - 910 - 920 - 930 - 940 - 950 - 960 - 970 - 980 - 990 - 1000

Notes: ____

____

____

____

Reg Reps: 10 - 20 - 30 - 40 - 50 - 60 - 70 - 80 - 90 - 100 - 110 - 120 - 130 - 140 - 150 - 160 - 170 - 180 - 190 - 200 - 210 - 220 - 230 - 240 - 250 - 260 - 270 - 280 - 290 - 300 - 310 - 320 - 330 - 340 - 350 - 360 - 370 - 380 - 390 - 400 - 410 - 420 - 430 - 440 - 450 - 460 - 470 - 480 - 490 - 500 - 510 - 520 - 530 - 540 - 550 - 560 - 570 - 580 - 590 - 600 - 610 - 620 - 630 - 640 - 650 - 660 - 670 - 680 - 690 - 700 - 710 - 720 - 730 - 740 - 750 - 760 - 770 - 780 - 790 - 800 - 810 - 820 - 830 - 840 - 850 - 860 - 870 - 880 - 890 - 900 - 910 - 920 - 930 - 940 - 950 - 960 - 970 - 980 - 990 - 1000

Notes: ____

____

____

______________________________ **End Date:** ____________

Kick: ______________________________ **Start Date:** __________

Kicking Leg: Left - Right - Alt. **Kicking Height: Low - Mid - High**

Notes: ____

____

____

Slow Reps: 10 - 20 - 30 - 40 - 50 - 60 - 70 - 80 - 90 - 100 - 110 - 120 - 130
140 - 150 - 160 - 170 - 180 - 190 - 200 - 210 - 220 - 230 - 240 - 250 - 260
270 - 280 - 290 - 300 - 310 - 320 - 330 - 340 - 350 - 360 - 370 - 380 - 390
400 - 410 - 420 - 430 - 440 - 450 - 460 - 470 - 480 - 490 - 500 - 510 - 520
530 - 540 - 550 - 560 - 570 - 580 - 590 - 600 - 610 - 620 - 630 - 640 - 650
660 - 670 - 680 - 690 - 700 - 710 - 720 - 730 - 740 - 750 - 760 - 770 - 780
790 - 800 - 810 - 820 - 830 - 840 - 850 - 860 - 870 - 880 - 890 - 900 - 910
920 - 930 - 940 - 950 - 960 - 970 - 980 - 990 - 1000

Reg Reps: 10 - 20 - 30 - 40 - 50 - 60 - 70 - 80 - 90 - 100 - 110 - 120 - 130
140 - 150 - 160 - 170 - 180 - 190 - 200 - 210 - 220 - 230 - 240 - 250 - 260
270 - 280 - 290 - 300 - 310 - 320 - 330 - 340 - 350 - 360 - 370 - 380 - 390
400 - 410 - 420 - 430 - 440 - 450 - 460 - 470 - 480 - 490 - 500 - 510 - 520
530 - 540 - 550 - 560 - 570 - 580 - 590 - 600 - 610 - 620 - 630 - 640 - 650
660 - 670 - 680 - 690 - 700 - 710 - 720 - 730 - 740 - 750 - 760 - 770 - 780
790 - 800 - 810 - 820 - 830 - 840 - 850 - 860 - 870 - 880 - 890 - 900 - 910
920 - 930 - 940 - 950 - 960 - 970 - 980 - 990 - 1000

Notes: ____

____

____

____

Reg Reps: 10 - 20 - 30 - 40 - 50 - 60 - 70 - 80 - 90 - 100 - 110 - 120 - 130
140 - 150 - 160 - 170 - 180 - 190 - 200 - 210 - 220 - 230 - 240 - 250 - 260
270 - 280 - 290 - 300 - 310 - 320 - 330 - 340 - 350 - 360 - 370 - 380 - 390
400 - 410 - 420 - 430 - 440 - 450 - 460 - 470 - 480 - 490 - 500 - 510 - 520
530 - 540 - 550 - 560 - 570 - 580 - 590 - 600 - 610 - 620 - 630 - 640 - 650
660 - 670 - 680 - 690 - 700 - 710 - 720 - 730 - 740 - 750 - 760 - 770 - 780
790 - 800 - 810 - 820 - 830 - 840 - 850 - 860 - 870 - 880 - 890 - 900 - 910
920 - 930 - 940 - 950 - 960 - 970 - 980 - 990 - 1000

Notes: ____

____

____

____

Reg Reps: 10 - 20 - 30 - 40 - 50 - 60 - 70 - 80 - 90 - 100 - 110 - 120 - 130
140 - 150 - 160 - 170 - 180 - 190 - 200 - 210 - 220 - 230 - 240 - 250 - 260
270 - 280 - 290 - 300 - 310 - 320 - 330 - 340 - 350 - 360 - 370 - 380 - 390
400 - 410 - 420 - 430 - 440 - 450 - 460 - 470 - 480 - 490 - 500 - 510 - 520
530 - 540 - 550 - 560 - 570 - 580 - 590 - 600 - 610 - 620 - 630 - 640 - 650
660 - 670 - 680 - 690 - 700 - 710 - 720 - 730 - 740 - 750 - 760 - 770 - 780
790 - 800 - 810 - 820 - 830 - 840 - 850 - 860 - 870 - 880 - 890 - 900 - 910
920 - 930 - 940 - 950 - 960 - 970 - 980 - 990 - 1000

Notes: ______________________________

____

____

____

Reg Reps: 10 - 20 - 30 - 40 - 50 - 60 - 70 - 80 - 90 - 100 - 110 - 120 - 130
140 - 150 - 160 - 170 - 180 - 190 - 200 - 210 - 220 - 230 - 240 - 250 - 260
270 - 280 - 290 - 300 - 310 - 320 - 330 - 340 - 350 - 360 - 370 - 380 - 390
400 - 410 - 420 - 430 - 440 - 450 - 460 - 470 - 480 - 490 - 500 - 510 - 520
530 - 540 - 550 - 560 - 570 - 580 - 590 - 600 - 610 - 620 - 630 - 640 - 650
660 - 670 - 680 - 690 - 700 - 710 - 720 - 730 - 740 - 750 - 760 - 770 - 780
790 - 800 - 810 - 820 - 830 - 840 - 850 - 860 - 870 - 880 - 890 - 900 - 910
920 - 930 - 940 - 950 - 960 - 970 - 980 - 990 - 1000

Notes: ______________________________

____

____

____

Reg Reps: 10 - 20 - 30 - 40 - 50 - 60 - 70 - 80 - 90 - 100 - 110 - 120 - 130
140 - 150 - 160 - 170 - 180 - 190 - 200 - 210 - 220 - 230 - 240 - 250 - 260
270 - 280 - 290 - 300 - 310 - 320 - 330 - 340 - 350 - 360 - 370 - 380 - 390
400 - 410 - 420 - 430 - 440 - 450 - 460 - 470 - 480 - 490 - 500 - 510 - 520
530 - 540 - 550 - 560 - 570 - 580 - 590 - 600 - 610 - 620 - 630 - 640 - 650
660 - 670 - 680 - 690 - 700 - 710 - 720 - 730 - 740 - 750 - 760 - 770 - 780
790 - 800 - 810 - 820 - 830 - 840 - 850 - 860 - 870 - 880 - 890 - 900 - 910
920 - 930 - 940 - 950 - 960 - 970 - 980 - 990 - 1000

Notes: ______________________________

____

____

______________________________ **End Date:** ____________

Kick: ______________________________ **Start Date:** __________

Kicking Leg: Left - Right - Alt. **Kicking Height: Low - Mid - High**

Notes: ____

____

____

Slow Reps: 10 - 20 - 30 - 40 - 50 - 60 - 70 - 80 - 90 - 100 - 110 - 120 - 130
140 - 150 - 160 - 170 - 180 - 190 - 200 - 210 - 220 - 230 - 240 - 250 - 260
270 - 280 - 290 - 300 - 310 - 320 - 330 - 340 - 350 - 360 - 370 - 380 - 390
400 - 410 - 420 - 430 - 440 - 450 - 460 - 470 - 480 - 490 - 500 - 510 - 520
530 - 540 - 550 - 560 - 570 - 580 - 590 - 600 - 610 - 620 - 630 - 640 - 650
660 - 670 - 680 - 690 - 700 - 710 - 720 - 730 - 740 - 750 - 760 - 770 - 780
790 - 800 - 810 - 820 - 830 - 840 - 850 - 860 - 870 - 880 - 890 - 900 - 910
920 - 930 - 940 - 950 - 960 - 970 - 980 - 990 - 1000

Reg Reps: 10 - 20 - 30 - 40 - 50 - 60 - 70 - 80 - 90 - 100 - 110 - 120 - 130
140 - 150 - 160 - 170 - 180 - 190 - 200 - 210 - 220 - 230 - 240 - 250 - 260
270 - 280 - 290 - 300 - 310 - 320 - 330 - 340 - 350 - 360 - 370 - 380 - 390
400 - 410 - 420 - 430 - 440 - 450 - 460 - 470 - 480 - 490 - 500 - 510 - 520
530 - 540 - 550 - 560 - 570 - 580 - 590 - 600 - 610 - 620 - 630 - 640 - 650
660 - 670 - 680 - 690 - 700 - 710 - 720 - 730 - 740 - 750 - 760 - 770 - 780
790 - 800 - 810 - 820 - 830 - 840 - 850 - 860 - 870 - 880 - 890 - 900 - 910
920 - 930 - 940 - 950 - 960 - 970 - 980 - 990 - 1000

Notes: ____

____

____

____

Reg Reps: 10 - 20 - 30 - 40 - 50 - 60 - 70 - 80 - 90 - 100 - 110 - 120 - 130
140 - 150 - 160 - 170 - 180 - 190 - 200 - 210 - 220 - 230 - 240 - 250 - 260
270 - 280 - 290 - 300 - 310 - 320 - 330 - 340 - 350 - 360 - 370 - 380 - 390
400 - 410 - 420 - 430 - 440 - 450 - 460 - 470 - 480 - 490 - 500 - 510 - 520
530 - 540 - 550 - 560 - 570 - 580 - 590 - 600 - 610 - 620 - 630 - 640 - 650
660 - 670 - 680 - 690 - 700 - 710 - 720 - 730 - 740 - 750 - 760 - 770 - 780
790 - 800 - 810 - 820 - 830 - 840 - 850 - 860 - 870 - 880 - 890 - 900 - 910
920 - 930 - 940 - 950 - 960 - 970 - 980 - 990 - 1000

Notes: ____

____

____

____

Reg Reps: 10 - 20 - 30 - 40 - 50 - 60 - 70 - 80 - 90 - 100 - 110 - 120 - 130
140 - 150 - 160 - 170 - 180 - 190 - 200 - 210 - 220 - 230 - 240 - 250 - 260
270 - 280 - 290 - 300 - 310 - 320 - 330 - 340 - 350 - 360 - 370 - 380 - 390
400 - 410 - 420 - 430 - 440 - 450 - 460 - 470 - 480 - 490 - 500 - 510 - 520
530 - 540 - 550 - 560 - 570 - 580 - 590 - 600 - 610 - 620 - 630 - 640 - 650
660 - 670 - 680 - 690 - 700 - 710 - 720 - 730 - 740 - 750 - 760 - 770 - 780
790 - 800 - 810 - 820 - 830 - 840 - 850 - 860 - 870 - 880 - 890 - 900 - 910
920 - 930 - 940 - 950 - 960 - 970 - 980 - 990 - 1000

Notes: ____

____

____

____

Reg Reps: 10 - 20 - 30 - 40 - 50 - 60 - 70 - 80 - 90 - 100 - 110 - 120 - 130
140 - 150 - 160 - 170 - 180 - 190 - 200 - 210 - 220 - 230 - 240 - 250 - 260
270 - 280 - 290 - 300 - 310 - 320 - 330 - 340 - 350 - 360 - 370 - 380 - 390
400 - 410 - 420 - 430 - 440 - 450 - 460 - 470 - 480 - 490 - 500 - 510 - 520
530 - 540 - 550 - 560 - 570 - 580 - 590 - 600 - 610 - 620 - 630 - 640 - 650
660 - 670 - 680 - 690 - 700 - 710 - 720 - 730 - 740 - 750 - 760 - 770 - 780
790 - 800 - 810 - 820 - 830 - 840 - 850 - 860 - 870 - 880 - 890 - 900 - 910
920 - 930 - 940 - 950 - 960 - 970 - 980 - 990 - 1000

Notes: ____

____

____

____

Reg Reps: 10 - 20 - 30 - 40 - 50 - 60 - 70 - 80 - 90 - 100 - 110 - 120 - 130
140 - 150 - 160 - 170 - 180 - 190 - 200 - 210 - 220 - 230 - 240 - 250 - 260
270 - 280 - 290 - 300 - 310 - 320 - 330 - 340 - 350 - 360 - 370 - 380 - 390
400 - 410 - 420 - 430 - 440 - 450 - 460 - 470 - 480 - 490 - 500 - 510 - 520
530 - 540 - 550 - 560 - 570 - 580 - 590 - 600 - 610 - 620 - 630 - 640 - 650
660 - 670 - 680 - 690 - 700 - 710 - 720 - 730 - 740 - 750 - 760 - 770 - 780
790 - 800 - 810 - 820 - 830 - 840 - 850 - 860 - 870 - 880 - 890 - 900 - 910
920 - 930 - 940 - 950 - 960 - 970 - 980 - 990 - 1000

Notes: ____

____

____

______________________________ **End Date:** ____________

Kick:____ **Start Date:** _________

Kicking Leg: Left - Right - Alt. **Kicking Height: Low - Mid - High**

Notes:____

Slow Reps: 10 - 20 - 30 - 40 - 50 - 60 - 70 - 80 - 90 - 100 - 110 - 120 - 130
140 - 150 - 160 - 170 - 180 - 190 - 200 - 210 - 220 - 230 - 240 - 250 - 260
270 - 280 - 290 - 300 - 310 - 320 - 330 - 340 - 350 - 360 - 370 - 380 - 390
400 - 410 - 420 - 430 - 440 - 450 - 460 - 470 - 480 - 490 - 500 - 510 - 520
530 - 540 - 550 - 560 - 570 - 580 - 590 - 600 - 610 - 620 - 630 - 640 - 650
660 - 670 - 680 - 690 - 700 - 710 - 720 - 730 - 740 - 750 - 760 - 770 - 780
790 - 800 - 810 - 820 - 830 - 840 - 850 - 860 - 870 - 880 - 890 - 900 - 910
920 - 930 - 940 - 950 - 960 - 970 - 980 - 990 - 1000

Reg Reps: 10 - 20 - 30 - 40 - 50 - 60 - 70 - 80 - 90 - 100 - 110 - 120 - 130
140 - 150 - 160 - 170 - 180 - 190 - 200 - 210 - 220 - 230 - 240 - 250 - 260
270 - 280 - 290 - 300 - 310 - 320 - 330 - 340 - 350 - 360 - 370 - 380 - 390
400 - 410 - 420 - 430 - 440 - 450 - 460 - 470 - 480 - 490 - 500 - 510 - 520
530 - 540 - 550 - 560 - 570 - 580 - 590 - 600 - 610 - 620 - 630 - 640 - 650
660 - 670 - 680 - 690 - 700 - 710 - 720 - 730 - 740 - 750 - 760 - 770 - 780
790 - 800 - 810 - 820 - 830 - 840 - 850 - 860 - 870 - 880 - 890 - 900 - 910
920 - 930 - 940 - 950 - 960 - 970 - 980 - 990 - 1000

Notes: ____

Reg Reps: 10 - 20 - 30 - 40 - 50 - 60 - 70 - 80 - 90 - 100 - 110 - 120 - 130
140 - 150 - 160 - 170 - 180 - 190 - 200 - 210 - 220 - 230 - 240 - 250 - 260
270 - 280 - 290 - 300 - 310 - 320 - 330 - 340 - 350 - 360 - 370 - 380 - 390
400 - 410 - 420 - 430 - 440 - 450 - 460 - 470 - 480 - 490 - 500 - 510 - 520
530 - 540 - 550 - 560 - 570 - 580 - 590 - 600 - 610 - 620 - 630 - 640 - 650
660 - 670 - 680 - 690 - 700 - 710 - 720 - 730 - 740 - 750 - 760 - 770 - 780
790 - 800 - 810 - 820 - 830 - 840 - 850 - 860 - 870 - 880 - 890 - 900 - 910
920 - 930 - 940 - 950 - 960 - 970 - 980 - 990 - 1000

Notes: ____

Reg Reps: 10 - 20 - 30 - 40 - 50 - 60 - 70 - 80 - 90 - 100 - 110 - 120 - 130
140 - 150 - 160 - 170 - 180 - 190 - 200 - 210 - 220 - 230 - 240 - 250 - 260
270 - 280 - 290 - 300 - 310 - 320 - 330 - 340 - 350 - 360 - 370 - 380 - 390
400 - 410 - 420 - 430 - 440 - 450 - 460 - 470 - 480 - 490 - 500 - 510 - 520
530 - 540 - 550 - 560 - 570 - 580 - 590 - 600 - 610 - 620 - 630 - 640 - 650
660 - 670 - 680 - 690 - 700 - 710 - 720 - 730 - 740 - 750 - 760 - 770 - 780
790 - 800 - 810 - 820 - 830 - 840 - 850 - 860 - 870 - 880 - 890 - 900 - 910
920 - 930 - 940 - 950 - 960 - 970 - 980 - 990 - 1000

Notes: ____

____

____

____

Reg Reps: 10 - 20 - 30 - 40 - 50 - 60 - 70 - 80 - 90 - 100 - 110 - 120 - 130
140 - 150 - 160 - 170 - 180 - 190 - 200 - 210 - 220 - 230 - 240 - 250 - 260
270 - 280 - 290 - 300 - 310 - 320 - 330 - 340 - 350 - 360 - 370 - 380 - 390
400 - 410 - 420 - 430 - 440 - 450 - 460 - 470 - 480 - 490 - 500 - 510 - 520
530 - 540 - 550 - 560 - 570 - 580 - 590 - 600 - 610 - 620 - 630 - 640 - 650
660 - 670 - 680 - 690 - 700 - 710 - 720 - 730 - 740 - 750 - 760 - 770 - 780
790 - 800 - 810 - 820 - 830 - 840 - 850 - 860 - 870 - 880 - 890 - 900 - 910
920 - 930 - 940 - 950 - 960 - 970 - 980 - 990 - 1000

Notes: ____

____

____

____

Reg Reps: 10 - 20 - 30 - 40 - 50 - 60 - 70 - 80 - 90 - 100 - 110 - 120 - 130
140 - 150 - 160 - 170 - 180 - 190 - 200 - 210 - 220 - 230 - 240 - 250 - 260
270 - 280 - 290 - 300 - 310 - 320 - 330 - 340 - 350 - 360 - 370 - 380 - 390
400 - 410 - 420 - 430 - 440 - 450 - 460 - 470 - 480 - 490 - 500 - 510 - 520
530 - 540 - 550 - 560 - 570 - 580 - 590 - 600 - 610 - 620 - 630 - 640 - 650
660 - 670 - 680 - 690 - 700 - 710 - 720 - 730 - 740 - 750 - 760 - 770 - 780
790 - 800 - 810 - 820 - 830 - 840 - 850 - 860 - 870 - 880 - 890 - 900 - 910
920 - 930 - 940 - 950 - 960 - 970 - 980 - 990 - 1000

Notes: ____

____

____

_________________________________ **End Date:** ____________

Kick:________________________________ **Start Date:** ________

Kicking Leg: Left - Right - Alt. **Kicking Height: Low - Mid - High**

Notes:____

____

____

Slow Reps: 10 - 20 - 30 - 40 - 50 - 60 - 70 - 80 - 90 - 100 - 110 - 120 - 130
140 - 150 - 160 - 170 - 180 - 190 - 200 - 210 - 220 - 230 - 240 - 250 - 260
270 - 280 - 290 - 300 - 310 - 320 - 330 - 340 - 350 - 360 - 370 - 380 - 390
400 - 410 - 420 - 430 - 440 - 450 - 460 - 470 - 480 - 490 - 500 - 510 - 520
530 - 540 - 550 - 560 - 570 - 580 - 590 - 600 - 610 - 620 - 630 - 640 - 650
660 - 670 - 680 - 690 - 700 - 710 - 720 - 730 - 740 - 750 - 760 - 770 - 780
790 - 800 - 810 - 820 - 830 - 840 - 850 - 860 - 870 - 880 - 890 - 900 - 910
920 - 930 - 940 - 950 - 960 - 970 - 980 - 990 - 1000

Reg Reps: 10 - 20 - 30 - 40 - 50 - 60 - 70 - 80 - 90 - 100 - 110 - 120 - 130
140 - 150 - 160 - 170 - 180 - 190 - 200 - 210 - 220 - 230 - 240 - 250 - 260
270 - 280 - 290 - 300 - 310 - 320 - 330 - 340 - 350 - 360 - 370 - 380 - 390
400 - 410 - 420 - 430 - 440 - 450 - 460 - 470 - 480 - 490 - 500 - 510 - 520
530 - 540 - 550 - 560 - 570 - 580 - 590 - 600 - 610 - 620 - 630 - 640 - 650
660 - 670 - 680 - 690 - 700 - 710 - 720 - 730 - 740 - 750 - 760 - 770 - 780
790 - 800 - 810 - 820 - 830 - 840 - 850 - 860 - 870 - 880 - 890 - 900 - 910
920 - 930 - 940 - 950 - 960 - 970 - 980 - 990 - 1000

Notes: ____

____

____

____

Reg Reps: 10 - 20 - 30 - 40 - 50 - 60 - 70 - 80 - 90 - 100 - 110 - 120 - 130
140 - 150 - 160 - 170 - 180 - 190 - 200 - 210 - 220 - 230 - 240 - 250 - 260
270 - 280 - 290 - 300 - 310 - 320 - 330 - 340 - 350 - 360 - 370 - 380 - 390
400 - 410 - 420 - 430 - 440 - 450 - 460 - 470 - 480 - 490 - 500 - 510 - 520
530 - 540 - 550 - 560 - 570 - 580 - 590 - 600 - 610 - 620 - 630 - 640 - 650
660 - 670 - 680 - 690 - 700 - 710 - 720 - 730 - 740 - 750 - 760 - 770 - 780
790 - 800 - 810 - 820 - 830 - 840 - 850 - 860 - 870 - 880 - 890 - 900 - 910
920 - 930 - 940 - 950 - 960 - 970 - 980 - 990 - 1000

Notes: ____

____

____

____

Reg Reps: 10 - 20 - 30 - 40 - 50 - 60 - 70 - 80 - 90 - 100 - 110 - 120 - 130 140 - 150 - 160 - 170 - 180 - 190 - 200 - 210 - 220 - 230 - 240 - 250 - 260 270 - 280 - 290 - 300 - 310 - 320 - 330 - 340 - 350 - 360 - 370 - 380 - 390 400 - 410 - 420 - 430 - 440 - 450 - 460 - 470 - 480 - 490 - 500 - 510 - 520 530 - 540 - 550 - 560 - 570 - 580 - 590 - 600 - 610 - 620 - 630 - 640 - 650 660 - 670 - 680 - 690 - 700 - 710 - 720 - 730 - 740 - 750 - 760 - 770 - 780 790 - 800 - 810 - 820 - 830 - 840 - 850 - 860 - 870 - 880 - 890 - 900 - 910 920 - 930 - 940 - 950 - 960 - 970 - 980 - 990 - 1000

Notes: ____

____

____

____

Reg Reps: 10 - 20 - 30 - 40 - 50 - 60 - 70 - 80 - 90 - 100 - 110 - 120 - 130 140 - 150 - 160 - 170 - 180 - 190 - 200 - 210 - 220 - 230 - 240 - 250 - 260 270 - 280 - 290 - 300 - 310 - 320 - 330 - 340 - 350 - 360 - 370 - 380 - 390 400 - 410 - 420 - 430 - 440 - 450 - 460 - 470 - 480 - 490 - 500 - 510 - 520 530 - 540 - 550 - 560 - 570 - 580 - 590 - 600 - 610 - 620 - 630 - 640 - 650 660 - 670 - 680 - 690 - 700 - 710 - 720 - 730 - 740 - 750 - 760 - 770 - 780 790 - 800 - 810 - 820 - 830 - 840 - 850 - 860 - 870 - 880 - 890 - 900 - 910 920 - 930 - 940 - 950 - 960 - 970 - 980 - 990 - 1000

Notes: ____

____

____

____

Reg Reps: 10 - 20 - 30 - 40 - 50 - 60 - 70 - 80 - 90 - 100 - 110 - 120 - 130 140 - 150 - 160 - 170 - 180 - 190 - 200 - 210 - 220 - 230 - 240 - 250 - 260 270 - 280 - 290 - 300 - 310 - 320 - 330 - 340 - 350 - 360 - 370 - 380 - 390 400 - 410 - 420 - 430 - 440 - 450 - 460 - 470 - 480 - 490 - 500 - 510 - 520 530 - 540 - 550 - 560 - 570 - 580 - 590 - 600 - 610 - 620 - 630 - 640 - 650 660 - 670 - 680 - 690 - 700 - 710 - 720 - 730 - 740 - 750 - 760 - 770 - 780 790 - 800 - 810 - 820 - 830 - 840 - 850 - 860 - 870 - 880 - 890 - 900 - 910 920 - 930 - 940 - 950 - 960 - 970 - 980 - 990 - 1000

Notes: ____

____

____

______________________________ **End Date:** ____________

Kick: ______________________________ **Start Date:** __________

Kicking Leg: Left - Right - Alt. **Kicking Height: Low - Mid - High**

Notes: ____

____

____

Slow Reps: 10 - 20 - 30 - 40 - 50 - 60 - 70 - 80 - 90 - 100 - 110 - 120 - 130
140 - 150 - 160 - 170 - 180 - 190 - 200 - 210 - 220 - 230 - 240 - 250 - 260
270 - 280 - 290 - 300 - 310 - 320 - 330 - 340 - 350 - 360 - 370 - 380 - 390
400 - 410 - 420 - 430 - 440 - 450 - 460 - 470 - 480 - 490 - 500 - 510 - 520
530 - 540 - 550 - 560 - 570 - 580 - 590 - 600 - 610 - 620 - 630 - 640 - 650
660 - 670 - 680 - 690 - 700 - 710 - 720 - 730 - 740 - 750 - 760 - 770 - 780
790 - 800 - 810 - 820 - 830 - 840 - 850 - 860 - 870 - 880 - 890 - 900 - 910
920 - 930 - 940 - 950 - 960 - 970 - 980 - 990 - 1000

Reg Reps: 10 - 20 - 30 - 40 - 50 - 60 - 70 - 80 - 90 - 100 - 110 - 120 - 130
140 - 150 - 160 - 170 - 180 - 190 - 200 - 210 - 220 - 230 - 240 - 250 - 260
270 - 280 - 290 - 300 - 310 - 320 - 330 - 340 - 350 - 360 - 370 - 380 - 390
400 - 410 - 420 - 430 - 440 - 450 - 460 - 470 - 480 - 490 - 500 - 510 - 520
530 - 540 - 550 - 560 - 570 - 580 - 590 - 600 - 610 - 620 - 630 - 640 - 650
660 - 670 - 680 - 690 - 700 - 710 - 720 - 730 - 740 - 750 - 760 - 770 - 780
790 - 800 - 810 - 820 - 830 - 840 - 850 - 860 - 870 - 880 - 890 - 900 - 910
920 - 930 - 940 - 950 - 960 - 970 - 980 - 990 - 1000

Notes: ____

____

____

____

Reg Reps: 10 - 20 - 30 - 40 - 50 - 60 - 70 - 80 - 90 - 100 - 110 - 120 - 130
140 - 150 - 160 - 170 - 180 - 190 - 200 - 210 - 220 - 230 - 240 - 250 - 260
270 - 280 - 290 - 300 - 310 - 320 - 330 - 340 - 350 - 360 - 370 - 380 - 390
400 - 410 - 420 - 430 - 440 - 450 - 460 - 470 - 480 - 490 - 500 - 510 - 520
530 - 540 - 550 - 560 - 570 - 580 - 590 - 600 - 610 - 620 - 630 - 640 - 650
660 - 670 - 680 - 690 - 700 - 710 - 720 - 730 - 740 - 750 - 760 - 770 - 780
790 - 800 - 810 - 820 - 830 - 840 - 850 - 860 - 870 - 880 - 890 - 900 - 910
920 - 930 - 940 - 950 - 960 - 970 - 980 - 990 - 1000

Notes: ____

____

____

____

Reg Reps: 10 - 20 - 30 - 40 - 50 - 60 - 70 - 80 - 90 - 100 - 110 - 120 - 130
140 - 150 - 160 - 170 - 180 - 190 - 200 - 210 - 220 - 230 - 240 - 250 - 260
270 - 280 - 290 - 300 - 310 - 320 - 330 - 340 - 350 - 360 - 370 - 380 - 390
400 - 410 - 420 - 430 - 440 - 450 - 460 - 470 - 480 - 490 - 500 - 510 - 520
530 - 540 - 550 - 560 - 570 - 580 - 590 - 600 - 610 - 620 - 630 - 640 - 650
660 - 670 - 680 - 690 - 700 - 710 - 720 - 730 - 740 - 750 - 760 - 770 - 780
790 - 800 - 810 - 820 - 830 - 840 - 850 - 860 - 870 - 880 - 890 - 900 - 910
920 - 930 - 940 - 950 - 960 - 970 - 980 - 990 - 1000

Notes: ____

____

____

____

Reg Reps: 10 - 20 - 30 - 40 - 50 - 60 - 70 - 80 - 90 - 100 - 110 - 120 - 130
140 - 150 - 160 - 170 - 180 - 190 - 200 - 210 - 220 - 230 - 240 - 250 - 260
270 - 280 - 290 - 300 - 310 - 320 - 330 - 340 - 350 - 360 - 370 - 380 - 390
400 - 410 - 420 - 430 - 440 - 450 - 460 - 470 - 480 - 490 - 500 - 510 - 520
530 - 540 - 550 - 560 - 570 - 580 - 590 - 600 - 610 - 620 - 630 - 640 - 650
660 - 670 - 680 - 690 - 700 - 710 - 720 - 730 - 740 - 750 - 760 - 770 - 780
790 - 800 - 810 - 820 - 830 - 840 - 850 - 860 - 870 - 880 - 890 - 900 - 910
920 - 930 - 940 - 950 - 960 - 970 - 980 - 990 - 1000

Notes: ____

____

____

____

Reg Reps: 10 - 20 - 30 - 40 - 50 - 60 - 70 - 80 - 90 - 100 - 110 - 120 - 130
140 - 150 - 160 - 170 - 180 - 190 - 200 - 210 - 220 - 230 - 240 - 250 - 260
270 - 280 - 290 - 300 - 310 - 320 - 330 - 340 - 350 - 360 - 370 - 380 - 390
400 - 410 - 420 - 430 - 440 - 450 - 460 - 470 - 480 - 490 - 500 - 510 - 520
530 - 540 - 550 - 560 - 570 - 580 - 590 - 600 - 610 - 620 - 630 - 640 - 650
660 - 670 - 680 - 690 - 700 - 710 - 720 - 730 - 740 - 750 - 760 - 770 - 780
790 - 800 - 810 - 820 - 830 - 840 - 850 - 860 - 870 - 880 - 890 - 900 - 910
920 - 930 - 940 - 950 - 960 - 970 - 980 - 990 - 1000

Notes: ____

____

____

______________________________ **End Date:** ____________

Kick:____________________________________ **Start Date:** __________

Kicking Leg: Left - Right - Alt. **Kicking Height: Low - Mid - High**

Notes:____

____

____

Slow Reps: 10 - 20 - 30 - 40 - 50 - 60 - 70 - 80 - 90 - 100 - 110 - 120 - 130
140 - 150 - 160 - 170 - 180 - 190 - 200 - 210 - 220 - 230 - 240 - 250 - 260
270 - 280 - 290 - 300 - 310 - 320 - 330 - 340 - 350 - 360 - 370 - 380 - 390
400 - 410 - 420 - 430 - 440 - 450 - 460 - 470 - 480 - 490 - 500 - 510 - 520
530 - 540 - 550 - 560 - 570 - 580 - 590 - 600 - 610 - 620 - 630 - 640 - 650
660 - 670 - 680 - 690 - 700 - 710 - 720 - 730 - 740 - 750 - 760 - 770 - 780
790 - 800 - 810 - 820 - 830 - 840 - 850 - 860 - 870 - 880 - 890 - 900 - 910
920 - 930 - 940 - 950 - 960 - 970 - 980 - 990 - 1000

Reg Reps: 10 - 20 - 30 - 40 - 50 - 60 - 70 - 80 - 90 - 100 - 110 - 120 - 130
140 - 150 - 160 - 170 - 180 - 190 - 200 - 210 - 220 - 230 - 240 - 250 - 260
270 - 280 - 290 - 300 - 310 - 320 - 330 - 340 - 350 - 360 - 370 - 380 - 390
400 - 410 - 420 - 430 - 440 - 450 - 460 - 470 - 480 - 490 - 500 - 510 - 520
530 - 540 - 550 - 560 - 570 - 580 - 590 - 600 - 610 - 620 - 630 - 640 - 650
660 - 670 - 680 - 690 - 700 - 710 - 720 - 730 - 740 - 750 - 760 - 770 - 780
790 - 800 - 810 - 820 - 830 - 840 - 850 - 860 - 870 - 880 - 890 - 900 - 910
920 - 930 - 940 - 950 - 960 - 970 - 980 - 990 - 1000

Notes: ____

____

____

____

Reg Reps: 10 - 20 - 30 - 40 - 50 - 60 - 70 - 80 - 90 - 100 - 110 - 120 - 130
140 - 150 - 160 - 170 - 180 - 190 - 200 - 210 - 220 - 230 - 240 - 250 - 260
270 - 280 - 290 - 300 - 310 - 320 - 330 - 340 - 350 - 360 - 370 - 380 - 390
400 - 410 - 420 - 430 - 440 - 450 - 460 - 470 - 480 - 490 - 500 - 510 - 520
530 - 540 - 550 - 560 - 570 - 580 - 590 - 600 - 610 - 620 - 630 - 640 - 650
660 - 670 - 680 - 690 - 700 - 710 - 720 - 730 - 740 - 750 - 760 - 770 - 780
790 - 800 - 810 - 820 - 830 - 840 - 850 - 860 - 870 - 880 - 890 - 900 - 910
920 - 930 - 940 - 950 - 960 - 970 - 980 - 990 - 1000

Notes: ____

____

____

____

Reg Reps: 10 - 20 - 30 - 40 - 50 - 60 - 70 - 80 - 90 - 100 - 110 - 120 - 130
140 - 150 - 160 - 170 - 180 - 190 - 200 - 210 - 220 - 230 - 240 - 250 - 260
270 - 280 - 290 - 300 - 310 - 320 - 330 - 340 - 350 - 360 - 370 - 380 - 390
400 - 410 - 420 - 430 - 440 - 450 - 460 - 470 - 480 - 490 - 500 - 510 - 520
530 - 540 - 550 - 560 - 570 - 580 - 590 - 600 - 610 - 620 - 630 - 640 - 650
660 - 670 - 680 - 690 - 700 - 710 - 720 - 730 - 740 - 750 - 760 - 770 - 780
790 - 800 - 810 - 820 - 830 - 840 - 850 - 860 - 870 - 880 - 890 - 900 - 910
920 - 930 - 940 - 950 - 960 - 970 - 980 - 990 - 1000

Notes: ____

____

____

____

Reg Reps: 10 - 20 - 30 - 40 - 50 - 60 - 70 - 80 - 90 - 100 - 110 - 120 - 130
140 - 150 - 160 - 170 - 180 - 190 - 200 - 210 - 220 - 230 - 240 - 250 - 260
270 - 280 - 290 - 300 - 310 - 320 - 330 - 340 - 350 - 360 - 370 - 380 - 390
400 - 410 - 420 - 430 - 440 - 450 - 460 - 470 - 480 - 490 - 500 - 510 - 520
530 - 540 - 550 - 560 - 570 - 580 - 590 - 600 - 610 - 620 - 630 - 640 - 650
660 - 670 - 680 - 690 - 700 - 710 - 720 - 730 - 740 - 750 - 760 - 770 - 780
790 - 800 - 810 - 820 - 830 - 840 - 850 - 860 - 870 - 880 - 890 - 900 - 910
920 - 930 - 940 - 950 - 960 - 970 - 980 - 990 - 1000

Notes: ____

____

____

____

Reg Reps: 10 - 20 - 30 - 40 - 50 - 60 - 70 - 80 - 90 - 100 - 110 - 120 - 130
140 - 150 - 160 - 170 - 180 - 190 - 200 - 210 - 220 - 230 - 240 - 250 - 260
270 - 280 - 290 - 300 - 310 - 320 - 330 - 340 - 350 - 360 - 370 - 380 - 390
400 - 410 - 420 - 430 - 440 - 450 - 460 - 470 - 480 - 490 - 500 - 510 - 520
530 - 540 - 550 - 560 - 570 - 580 - 590 - 600 - 610 - 620 - 630 - 640 - 650
660 - 670 - 680 - 690 - 700 - 710 - 720 - 730 - 740 - 750 - 760 - 770 - 780
790 - 800 - 810 - 820 - 830 - 840 - 850 - 860 - 870 - 880 - 890 - 900 - 910
920 - 930 - 940 - 950 - 960 - 970 - 980 - 990 - 1000

Notes: ____

____

____

______________________________ **End Date:** ____________

Kick:____ **Start Date:** __________

Kicking Leg: Left - Right - Alt. **Kicking Height: Low - Mid - High**

Notes:____

____

____

Slow Reps: 10 - 20 - 30 - 40 - 50 - 60 - 70 - 80 - 90 - 100 - 110 - 120 - 130
140 - 150 - 160 - 170 - 180 - 190 - 200 - 210 - 220 - 230 - 240 - 250 - 260
270 - 280 - 290 - 300 - 310 - 320 - 330 - 340 - 350 - 360 - 370 - 380 - 390
400 - 410 - 420 - 430 - 440 - 450 - 460 - 470 - 480 - 490 - 500 - 510 - 520
530 - 540 - 550 - 560 - 570 - 580 - 590 - 600 - 610 - 620 - 630 - 640 - 650
660 - 670 - 680 - 690 - 700 - 710 - 720 - 730 - 740 - 750 - 760 - 770 - 780
790 - 800 - 810 - 820 - 830 - 840 - 850 - 860 - 870 - 880 - 890 - 900 - 910
920 - 930 - 940 - 950 - 960 - 970 - 980 - 990 - 1000

Reg Reps: 10 - 20 - 30 - 40 - 50 - 60 - 70 - 80 - 90 - 100 - 110 - 120 - 130
140 - 150 - 160 - 170 - 180 - 190 - 200 - 210 - 220 - 230 - 240 - 250 - 260
270 - 280 - 290 - 300 - 310 - 320 - 330 - 340 - 350 - 360 - 370 - 380 - 390
400 - 410 - 420 - 430 - 440 - 450 - 460 - 470 - 480 - 490 - 500 - 510 - 520
530 - 540 - 550 - 560 - 570 - 580 - 590 - 600 - 610 - 620 - 630 - 640 - 650
660 - 670 - 680 - 690 - 700 - 710 - 720 - 730 - 740 - 750 - 760 - 770 - 780
790 - 800 - 810 - 820 - 830 - 840 - 850 - 860 - 870 - 880 - 890 - 900 - 910
920 - 930 - 940 - 950 - 960 - 970 - 980 - 990 - 1000

Notes: ____

____

____

____

Reg Reps: 10 - 20 - 30 - 40 - 50 - 60 - 70 - 80 - 90 - 100 - 110 - 120 - 130
140 - 150 - 160 - 170 - 180 - 190 - 200 - 210 - 220 - 230 - 240 - 250 - 260
270 - 280 - 290 - 300 - 310 - 320 - 330 - 340 - 350 - 360 - 370 - 380 - 390
400 - 410 - 420 - 430 - 440 - 450 - 460 - 470 - 480 - 490 - 500 - 510 - 520
530 - 540 - 550 - 560 - 570 - 580 - 590 - 600 - 610 - 620 - 630 - 640 - 650
660 - 670 - 680 - 690 - 700 - 710 - 720 - 730 - 740 - 750 - 760 - 770 - 780
790 - 800 - 810 - 820 - 830 - 840 - 850 - 860 - 870 - 880 - 890 - 900 - 910
920 - 930 - 940 - 950 - 960 - 970 - 980 - 990 - 1000

Notes: ____

____

____

____

Reg Reps: 10 - 20 - 30 - 40 - 50 - 60 - 70 - 80 - 90 - 100 - 110 - 120 - 130
140 - 150 - 160 - 170 - 180 - 190 - 200 - 210 - 220 - 230 - 240 - 250 - 260
270 - 280 - 290 - 300 - 310 - 320 - 330 - 340 - 350 - 360 - 370 - 380 - 390
400 - 410 - 420 - 430 - 440 - 450 - 460 - 470 - 480 - 490 - 500 - 510 - 520
530 - 540 - 550 - 560 - 570 - 580 - 590 - 600 - 610 - 620 - 630 - 640 - 650
660 - 670 - 680 - 690 - 700 - 710 - 720 - 730 - 740 - 750 - 760 - 770 - 780
790 - 800 - 810 - 820 - 830 - 840 - 850 - 860 - 870 - 880 - 890 - 900 - 910
920 - 930 - 940 - 950 - 960 - 970 - 980 - 990 - 1000

Notes: ____

____

____

____

Reg Reps: 10 - 20 - 30 - 40 - 50 - 60 - 70 - 80 - 90 - 100 - 110 - 120 - 130
140 - 150 - 160 - 170 - 180 - 190 - 200 - 210 - 220 - 230 - 240 - 250 - 260
270 - 280 - 290 - 300 - 310 - 320 - 330 - 340 - 350 - 360 - 370 - 380 - 390
400 - 410 - 420 - 430 - 440 - 450 - 460 - 470 - 480 - 490 - 500 - 510 - 520
530 - 540 - 550 - 560 - 570 - 580 - 590 - 600 - 610 - 620 - 630 - 640 - 650
660 - 670 - 680 - 690 - 700 - 710 - 720 - 730 - 740 - 750 - 760 - 770 - 780
790 - 800 - 810 - 820 - 830 - 840 - 850 - 860 - 870 - 880 - 890 - 900 - 910
920 - 930 - 940 - 950 - 960 - 970 - 980 - 990 - 1000

Notes: ____

____

____

____

Reg Reps: 10 - 20 - 30 - 40 - 50 - 60 - 70 - 80 - 90 - 100 - 110 - 120 - 130
140 - 150 - 160 - 170 - 180 - 190 - 200 - 210 - 220 - 230 - 240 - 250 - 260
270 - 280 - 290 - 300 - 310 - 320 - 330 - 340 - 350 - 360 - 370 - 380 - 390
400 - 410 - 420 - 430 - 440 - 450 - 460 - 470 - 480 - 490 - 500 - 510 - 520
530 - 540 - 550 - 560 - 570 - 580 - 590 - 600 - 610 - 620 - 630 - 640 - 650
660 - 670 - 680 - 690 - 700 - 710 - 720 - 730 - 740 - 750 - 760 - 770 - 780
790 - 800 - 810 - 820 - 830 - 840 - 850 - 860 - 870 - 880 - 890 - 900 - 910
920 - 930 - 940 - 950 - 960 - 970 - 980 - 990 - 1000

Notes: ____

____

____

______________________________ **End Date:** ____________

Kick: ________________________________ **Start Date:** ________

Kicking Leg: Left - Right - Alt. **Kicking Height: Low - Mid - High**

Notes: ____

____

____

Slow Reps: 10 - 20 - 30 - 40 - 50 - 60 - 70 - 80 - 90 - 100 - 110 - 120 - 130
140 - 150 - 160 - 170 - 180 - 190 - 200 - 210 - 220 - 230 - 240 - 250 - 260
270 - 280 - 290 - 300 - 310 - 320 - 330 - 340 - 350 - 360 - 370 - 380 - 390
400 - 410 - 420 - 430 - 440 - 450 - 460 - 470 - 480 - 490 - 500 - 510 - 520
530 - 540 - 550 - 560 - 570 - 580 - 590 - 600 - 610 - 620 - 630 - 640 - 650
660 - 670 - 680 - 690 - 700 - 710 - 720 - 730 - 740 - 750 - 760 - 770 - 780
790 - 800 - 810 - 820 - 830 - 840 - 850 - 860 - 870 - 880 - 890 - 900 - 910
920 - 930 - 940 - 950 - 960 - 970 - 980 - 990 - 1000

Reg Reps: 10 - 20 - 30 - 40 - 50 - 60 - 70 - 80 - 90 - 100 - 110 - 120 - 130
140 - 150 - 160 - 170 - 180 - 190 - 200 - 210 - 220 - 230 - 240 - 250 - 260
270 - 280 - 290 - 300 - 310 - 320 - 330 - 340 - 350 - 360 - 370 - 380 - 390
400 - 410 - 420 - 430 - 440 - 450 - 460 - 470 - 480 - 490 - 500 - 510 - 520
530 - 540 - 550 - 560 - 570 - 580 - 590 - 600 - 610 - 620 - 630 - 640 - 650
660 - 670 - 680 - 690 - 700 - 710 - 720 - 730 - 740 - 750 - 760 - 770 - 780
790 - 800 - 810 - 820 - 830 - 840 - 850 - 860 - 870 - 880 - 890 - 900 - 910
920 - 930 - 940 - 950 - 960 - 970 - 980 - 990 - 1000

Notes: ____

____

____

____

Reg Reps: 10 - 20 - 30 - 40 - 50 - 60 - 70 - 80 - 90 - 100 - 110 - 120 - 130
140 - 150 - 160 - 170 - 180 - 190 - 200 - 210 - 220 - 230 - 240 - 250 - 260
270 - 280 - 290 - 300 - 310 - 320 - 330 - 340 - 350 - 360 - 370 - 380 - 390
400 - 410 - 420 - 430 - 440 - 450 - 460 - 470 - 480 - 490 - 500 - 510 - 520
530 - 540 - 550 - 560 - 570 - 580 - 590 - 600 - 610 - 620 - 630 - 640 - 650
660 - 670 - 680 - 690 - 700 - 710 - 720 - 730 - 740 - 750 - 760 - 770 - 780
790 - 800 - 810 - 820 - 830 - 840 - 850 - 860 - 870 - 880 - 890 - 900 - 910
920 - 930 - 940 - 950 - 960 - 970 - 980 - 990 - 1000

Notes: ____

____

____

____

Reg Reps: 10 - 20 - 30 - 40 - 50 - 60 - 70 - 80 - 90 - 100 - 110 - 120 - 130
140 - 150 - 160 - 170 - 180 - 190 - 200 - 210 - 220 - 230 - 240 - 250 - 260
270 - 280 - 290 - 300 - 310 - 320 - 330 - 340 - 350 - 360 - 370 - 380 - 390
400 - 410 - 420 - 430 - 440 - 450 - 460 - 470 - 480 - 490 - 500 - 510 - 520
530 - 540 - 550 - 560 - 570 - 580 - 590 - 600 - 610 - 620 - 630 - 640 - 650
660 - 670 - 680 - 690 - 700 - 710 - 720 - 730 - 740 - 750 - 760 - 770 - 780
790 - 800 - 810 - 820 - 830 - 840 - 850 - 860 - 870 - 880 - 890 - 900 - 910
920 - 930 - 940 - 950 - 960 - 970 - 980 - 990 - 1000

Notes: ______________________________

____

____

____

Reg Reps: 10 - 20 - 30 - 40 - 50 - 60 - 70 - 80 - 90 - 100 - 110 - 120 - 130
140 - 150 - 160 - 170 - 180 - 190 - 200 - 210 - 220 - 230 - 240 - 250 - 260
270 - 280 - 290 - 300 - 310 - 320 - 330 - 340 - 350 - 360 - 370 - 380 - 390
400 - 410 - 420 - 430 - 440 - 450 - 460 - 470 - 480 - 490 - 500 - 510 - 520
530 - 540 - 550 - 560 - 570 - 580 - 590 - 600 - 610 - 620 - 630 - 640 - 650
660 - 670 - 680 - 690 - 700 - 710 - 720 - 730 - 740 - 750 - 760 - 770 - 780
790 - 800 - 810 - 820 - 830 - 840 - 850 - 860 - 870 - 880 - 890 - 900 - 910
920 - 930 - 940 - 950 - 960 - 970 - 980 - 990 - 1000

Notes: ______________________________

____

____

____

Reg Reps: 10 - 20 - 30 - 40 - 50 - 60 - 70 - 80 - 90 - 100 - 110 - 120 - 130
140 - 150 - 160 - 170 - 180 - 190 - 200 - 210 - 220 - 230 - 240 - 250 - 260
270 - 280 - 290 - 300 - 310 - 320 - 330 - 340 - 350 - 360 - 370 - 380 - 390
400 - 410 - 420 - 430 - 440 - 450 - 460 - 470 - 480 - 490 - 500 - 510 - 520
530 - 540 - 550 - 560 - 570 - 580 - 590 - 600 - 610 - 620 - 630 - 640 - 650
660 - 670 - 680 - 690 - 700 - 710 - 720 - 730 - 740 - 750 - 760 - 770 - 780
790 - 800 - 810 - 820 - 830 - 840 - 850 - 860 - 870 - 880 - 890 - 900 - 910
920 - 930 - 940 - 950 - 960 - 970 - 980 - 990 - 1000

Notes: ______________________________

____

____

______________________________ **End Date:** ____________

Kick:____ **Start Date:** __________

Kicking Leg: Left - Right - Alt. **Kicking Height: Low - Mid - High**

Notes:____

____

____

Slow Reps: 10 - 20 - 30 - 40 - 50 - 60 - 70 - 80 - 90 - 100 - 110 - 120 - 130
140 - 150 - 160 - 170 - 180 - 190 - 200 - 210 - 220 - 230 - 240 - 250 - 260
270 - 280 - 290 - 300 - 310 - 320 - 330 - 340 - 350 - 360 - 370 - 380 - 390
400 - 410 - 420 - 430 - 440 - 450 - 460 - 470 - 480 - 490 - 500 - 510 - 520
530 - 540 - 550 - 560 - 570 - 580 - 590 - 600 - 610 - 620 - 630 - 640 - 650
660 - 670 - 680 - 690 - 700 - 710 - 720 - 730 - 740 - 750 - 760 - 770 - 780
790 - 800 - 810 - 820 - 830 - 840 - 850 - 860 - 870 - 880 - 890 - 900 - 910
920 - 930 - 940 - 950 - 960 - 970 - 980 - 990 - 1000

Reg Reps: 10 - 20 - 30 - 40 - 50 - 60 - 70 - 80 - 90 - 100 - 110 - 120 - 130
140 - 150 - 160 - 170 - 180 - 190 - 200 - 210 - 220 - 230 - 240 - 250 - 260
270 - 280 - 290 - 300 - 310 - 320 - 330 - 340 - 350 - 360 - 370 - 380 - 390
400 - 410 - 420 - 430 - 440 - 450 - 460 - 470 - 480 - 490 - 500 - 510 - 520
530 - 540 - 550 - 560 - 570 - 580 - 590 - 600 - 610 - 620 - 630 - 640 - 650
660 - 670 - 680 - 690 - 700 - 710 - 720 - 730 - 740 - 750 - 760 - 770 - 780
790 - 800 - 810 - 820 - 830 - 840 - 850 - 860 - 870 - 880 - 890 - 900 - 910
920 - 930 - 940 - 950 - 960 - 970 - 980 - 990 - 1000

Notes: ____

____

____

____

Reg Reps: 10 - 20 - 30 - 40 - 50 - 60 - 70 - 80 - 90 - 100 - 110 - 120 - 130
140 - 150 - 160 - 170 - 180 - 190 - 200 - 210 - 220 - 230 - 240 - 250 - 260
270 - 280 - 290 - 300 - 310 - 320 - 330 - 340 - 350 - 360 - 370 - 380 - 390
400 - 410 - 420 - 430 - 440 - 450 - 460 - 470 - 480 - 490 - 500 - 510 - 520
530 - 540 - 550 - 560 - 570 - 580 - 590 - 600 - 610 - 620 - 630 - 640 - 650
660 - 670 - 680 - 690 - 700 - 710 - 720 - 730 - 740 - 750 - 760 - 770 - 780
790 - 800 - 810 - 820 - 830 - 840 - 850 - 860 - 870 - 880 - 890 - 900 - 910
920 - 930 - 940 - 950 - 960 - 970 - 980 - 990 - 1000

Notes: ____

____

____

____

Reg Reps: 10 - 20 - 30 - 40 - 50 - 60 - 70 - 80 - 90 - 100 - 110 - 120 - 130 - 140 - 150 - 160 - 170 - 180 - 190 - 200 - 210 - 220 - 230 - 240 - 250 - 260 - 270 - 280 - 290 - 300 - 310 - 320 - 330 - 340 - 350 - 360 - 370 - 380 - 390 - 400 - 410 - 420 - 430 - 440 - 450 - 460 - 470 - 480 - 490 - 500 - 510 - 520 - 530 - 540 - 550 - 560 - 570 - 580 - 590 - 600 - 610 - 620 - 630 - 640 - 650 - 660 - 670 - 680 - 690 - 700 - 710 - 720 - 730 - 740 - 750 - 760 - 770 - 780 - 790 - 800 - 810 - 820 - 830 - 840 - 850 - 860 - 870 - 880 - 890 - 900 - 910 - 920 - 930 - 940 - 950 - 960 - 970 - 980 - 990 - 1000

Notes: ______________________________

____

____

____

Reg Reps: 10 - 20 - 30 - 40 - 50 - 60 - 70 - 80 - 90 - 100 - 110 - 120 - 130 - 140 - 150 - 160 - 170 - 180 - 190 - 200 - 210 - 220 - 230 - 240 - 250 - 260 - 270 - 280 - 290 - 300 - 310 - 320 - 330 - 340 - 350 - 360 - 370 - 380 - 390 - 400 - 410 - 420 - 430 - 440 - 450 - 460 - 470 - 480 - 490 - 500 - 510 - 520 - 530 - 540 - 550 - 560 - 570 - 580 - 590 - 600 - 610 - 620 - 630 - 640 - 650 - 660 - 670 - 680 - 690 - 700 - 710 - 720 - 730 - 740 - 750 - 760 - 770 - 780 - 790 - 800 - 810 - 820 - 830 - 840 - 850 - 860 - 870 - 880 - 890 - 900 - 910 - 920 - 930 - 940 - 950 - 960 - 970 - 980 - 990 - 1000

Notes: ______________________________

____

____

____

Reg Reps: 10 - 20 - 30 - 40 - 50 - 60 - 70 - 80 - 90 - 100 - 110 - 120 - 130 - 140 - 150 - 160 - 170 - 180 - 190 - 200 - 210 - 220 - 230 - 240 - 250 - 260 - 270 - 280 - 290 - 300 - 310 - 320 - 330 - 340 - 350 - 360 - 370 - 380 - 390 - 400 - 410 - 420 - 430 - 440 - 450 - 460 - 470 - 480 - 490 - 500 - 510 - 520 - 530 - 540 - 550 - 560 - 570 - 580 - 590 - 600 - 610 - 620 - 630 - 640 - 650 - 660 - 670 - 680 - 690 - 700 - 710 - 720 - 730 - 740 - 750 - 760 - 770 - 780 - 790 - 800 - 810 - 820 - 830 - 840 - 850 - 860 - 870 - 880 - 890 - 900 - 910 - 920 - 930 - 940 - 950 - 960 - 970 - 980 - 990 - 1000

Notes: ______________________________

____

____

______________________________ **End Date:** ____________

Kick:______________________________ **Start Date:** __________

Kicking Leg: Left - Right - Alt. **Kicking Height: Low - Mid - High**

Notes:____

____

____

Slow Reps: 10 - 20 - 30 - 40 - 50 - 60 - 70 - 80 - 90 - 100 - 110 - 120 - 130
140 - 150 - 160 - 170 - 180 - 190 - 200 - 210 - 220 - 230 - 240 - 250 - 260
270 - 280 - 290 - 300 - 310 - 320 - 330 - 340 - 350 - 360 - 370 - 380 - 390
400 - 410 - 420 - 430 - 440 - 450 - 460 - 470 - 480 - 490 - 500 - 510 - 520
530 - 540 - 550 - 560 - 570 - 580 - 590 - 600 - 610 - 620 - 630 - 640 - 650
660 - 670 - 680 - 690 - 700 - 710 - 720 - 730 - 740 - 750 - 760 - 770 - 780
790 - 800 - 810 - 820 - 830 - 840 - 850 - 860 - 870 - 880 - 890 - 900 - 910
920 - 930 - 940 - 950 - 960 - 970 - 980 - 990 - 1000

Reg Reps: 10 - 20 - 30 - 40 - 50 - 60 - 70 - 80 - 90 - 100 - 110 - 120 - 130
140 - 150 - 160 - 170 - 180 - 190 - 200 - 210 - 220 - 230 - 240 - 250 - 260
270 - 280 - 290 - 300 - 310 - 320 - 330 - 340 - 350 - 360 - 370 - 380 - 390
400 - 410 - 420 - 430 - 440 - 450 - 460 - 470 - 480 - 490 - 500 - 510 - 520
530 - 540 - 550 - 560 - 570 - 580 - 590 - 600 - 610 - 620 - 630 - 640 - 650
660 - 670 - 680 - 690 - 700 - 710 - 720 - 730 - 740 - 750 - 760 - 770 - 780
790 - 800 - 810 - 820 - 830 - 840 - 850 - 860 - 870 - 880 - 890 - 900 - 910
920 - 930 - 940 - 950 - 960 - 970 - 980 - 990 - 1000

Notes: ____

____

____

____

Reg Reps: 10 - 20 - 30 - 40 - 50 - 60 - 70 - 80 - 90 - 100 - 110 - 120 - 130
140 - 150 - 160 - 170 - 180 - 190 - 200 - 210 - 220 - 230 - 240 - 250 - 260
270 - 280 - 290 - 300 - 310 - 320 - 330 - 340 - 350 - 360 - 370 - 380 - 390
400 - 410 - 420 - 430 - 440 - 450 - 460 - 470 - 480 - 490 - 500 - 510 - 520
530 - 540 - 550 - 560 - 570 - 580 - 590 - 600 - 610 - 620 - 630 - 640 - 650
660 - 670 - 680 - 690 - 700 - 710 - 720 - 730 - 740 - 750 - 760 - 770 - 780
790 - 800 - 810 - 820 - 830 - 840 - 850 - 860 - 870 - 880 - 890 - 900 - 910
920 - 930 - 940 - 950 - 960 - 970 - 980 - 990 - 1000

Notes: ____

____

____

____

Reg Reps: 10 - 20 - 30 - 40 - 50 - 60 - 70 - 80 - 90 - 100 - 110 - 120 - 130
140 - 150 - 160 - 170 - 180 - 190 - 200 - 210 - 220 - 230 - 240 - 250 - 260
270 - 280 - 290 - 300 - 310 - 320 - 330 - 340 - 350 - 360 - 370 - 380 - 390
400 - 410 - 420 - 430 - 440 - 450 - 460 - 470 - 480 - 490 - 500 - 510 - 520
530 - 540 - 550 - 560 - 570 - 580 - 590 - 600 - 610 - 620 - 630 - 640 - 650
660 - 670 - 680 - 690 - 700 - 710 - 720 - 730 - 740 - 750 - 760 - 770 - 780
790 - 800 - 810 - 820 - 830 - 840 - 850 - 860 - 870 - 880 - 890 - 900 - 910
920 - 930 - 940 - 950 - 960 - 970 - 980 - 990 - 1000

Notes: ____

____

____

____

Reg Reps: 10 - 20 - 30 - 40 - 50 - 60 - 70 - 80 - 90 - 100 - 110 - 120 - 130
140 - 150 - 160 - 170 - 180 - 190 - 200 - 210 - 220 - 230 - 240 - 250 - 260
270 - 280 - 290 - 300 - 310 - 320 - 330 - 340 - 350 - 360 - 370 - 380 - 390
400 - 410 - 420 - 430 - 440 - 450 - 460 - 470 - 480 - 490 - 500 - 510 - 520
530 - 540 - 550 - 560 - 570 - 580 - 590 - 600 - 610 - 620 - 630 - 640 - 650
660 - 670 - 680 - 690 - 700 - 710 - 720 - 730 - 740 - 750 - 760 - 770 - 780
790 - 800 - 810 - 820 - 830 - 840 - 850 - 860 - 870 - 880 - 890 - 900 - 910
920 - 930 - 940 - 950 - 960 - 970 - 980 - 990 - 1000

Notes: ____

____

____

____

Reg Reps: 10 - 20 - 30 - 40 - 50 - 60 - 70 - 80 - 90 - 100 - 110 - 120 - 130
140 - 150 - 160 - 170 - 180 - 190 - 200 - 210 - 220 - 230 - 240 - 250 - 260
270 - 280 - 290 - 300 - 310 - 320 - 330 - 340 - 350 - 360 - 370 - 380 - 390
400 - 410 - 420 - 430 - 440 - 450 - 460 - 470 - 480 - 490 - 500 - 510 - 520
530 - 540 - 550 - 560 - 570 - 580 - 590 - 600 - 610 - 620 - 630 - 640 - 650
660 - 670 - 680 - 690 - 700 - 710 - 720 - 730 - 740 - 750 - 760 - 770 - 780
790 - 800 - 810 - 820 - 830 - 840 - 850 - 860 - 870 - 880 - 890 - 900 - 910
920 - 930 - 940 - 950 - 960 - 970 - 980 - 990 - 1000

Notes: ____

____

____

______________________________ **End Date:** ____________

Kick: ______________________________ **Start Date:** ________

Kicking Leg: Left - Right - Alt. **Kicking Height: Low - Mid - High**

Notes: ____

____

____

Slow Reps: 10 - 20 - 30 - 40 - 50 - 60 - 70 - 80 - 90 - 100 - 110 - 120 - 130
140 - 150 - 160 - 170 - 180 - 190 - 200 - 210 - 220 - 230 - 240 - 250 - 260
270 - 280 - 290 - 300 - 310 - 320 - 330 - 340 - 350 - 360 - 370 - 380 - 390
400 - 410 - 420 - 430 - 440 - 450 - 460 - 470 - 480 - 490 - 500 - 510 - 520
530 - 540 - 550 - 560 - 570 - 580 - 590 - 600 - 610 - 620 - 630 - 640 - 650
660 - 670 - 680 - 690 - 700 - 710 - 720 - 730 - 740 - 750 - 760 - 770 - 780
790 - 800 - 810 - 820 - 830 - 840 - 850 - 860 - 870 - 880 - 890 - 900 - 910
920 - 930 - 940 - 950 - 960 - 970 - 980 - 990 - 1000

Reg Reps: 10 - 20 - 30 - 40 - 50 - 60 - 70 - 80 - 90 - 100 - 110 - 120 - 130
140 - 150 - 160 - 170 - 180 - 190 - 200 - 210 - 220 - 230 - 240 - 250 - 260
270 - 280 - 290 - 300 - 310 - 320 - 330 - 340 - 350 - 360 - 370 - 380 - 390
400 - 410 - 420 - 430 - 440 - 450 - 460 - 470 - 480 - 490 - 500 - 510 - 520
530 - 540 - 550 - 560 - 570 - 580 - 590 - 600 - 610 - 620 - 630 - 640 - 650
660 - 670 - 680 - 690 - 700 - 710 - 720 - 730 - 740 - 750 - 760 - 770 - 780
790 - 800 - 810 - 820 - 830 - 840 - 850 - 860 - 870 - 880 - 890 - 900 - 910
920 - 930 - 940 - 950 - 960 - 970 - 980 - 990 - 1000

Notes: ____

____

____

____

Reg Reps: 10 - 20 - 30 - 40 - 50 - 60 - 70 - 80 - 90 - 100 - 110 - 120 - 130
140 - 150 - 160 - 170 - 180 - 190 - 200 - 210 - 220 - 230 - 240 - 250 - 260
270 - 280 - 290 - 300 - 310 - 320 - 330 - 340 - 350 - 360 - 370 - 380 - 390
400 - 410 - 420 - 430 - 440 - 450 - 460 - 470 - 480 - 490 - 500 - 510 - 520
530 - 540 - 550 - 560 - 570 - 580 - 590 - 600 - 610 - 620 - 630 - 640 - 650
660 - 670 - 680 - 690 - 700 - 710 - 720 - 730 - 740 - 750 - 760 - 770 - 780
790 - 800 - 810 - 820 - 830 - 840 - 850 - 860 - 870 - 880 - 890 - 900 - 910
920 - 930 - 940 - 950 - 960 - 970 - 980 - 990 - 1000

Notes: ____

____

____

____

Reg Reps: 10 - 20 - 30 - 40 - 50 - 60 - 70 - 80 - 90 - 100 - 110 - 120 - 130 - 140 - 150 - 160 - 170 - 180 - 190 - 200 - 210 - 220 - 230 - 240 - 250 - 260 - 270 - 280 - 290 - 300 - 310 - 320 - 330 - 340 - 350 - 360 - 370 - 380 - 390 - 400 - 410 - 420 - 430 - 440 - 450 - 460 - 470 - 480 - 490 - 500 - 510 - 520 - 530 - 540 - 550 - 560 - 570 - 580 - 590 - 600 - 610 - 620 - 630 - 640 - 650 - 660 - 670 - 680 - 690 - 700 - 710 - 720 - 730 - 740 - 750 - 760 - 770 - 780 - 790 - 800 - 810 - 820 - 830 - 840 - 850 - 860 - 870 - 880 - 890 - 900 - 910 - 920 - 930 - 940 - 950 - 960 - 970 - 980 - 990 - 1000

Notes: ____

____

____

____

Reg Reps: 10 - 20 - 30 - 40 - 50 - 60 - 70 - 80 - 90 - 100 - 110 - 120 - 130 - 140 - 150 - 160 - 170 - 180 - 190 - 200 - 210 - 220 - 230 - 240 - 250 - 260 - 270 - 280 - 290 - 300 - 310 - 320 - 330 - 340 - 350 - 360 - 370 - 380 - 390 - 400 - 410 - 420 - 430 - 440 - 450 - 460 - 470 - 480 - 490 - 500 - 510 - 520 - 530 - 540 - 550 - 560 - 570 - 580 - 590 - 600 - 610 - 620 - 630 - 640 - 650 - 660 - 670 - 680 - 690 - 700 - 710 - 720 - 730 - 740 - 750 - 760 - 770 - 780 - 790 - 800 - 810 - 820 - 830 - 840 - 850 - 860 - 870 - 880 - 890 - 900 - 910 - 920 - 930 - 940 - 950 - 960 - 970 - 980 - 990 - 1000

Notes: ____

____

____

____

Reg Reps: 10 - 20 - 30 - 40 - 50 - 60 - 70 - 80 - 90 - 100 - 110 - 120 - 130 - 140 - 150 - 160 - 170 - 180 - 190 - 200 - 210 - 220 - 230 - 240 - 250 - 260 - 270 - 280 - 290 - 300 - 310 - 320 - 330 - 340 - 350 - 360 - 370 - 380 - 390 - 400 - 410 - 420 - 430 - 440 - 450 - 460 - 470 - 480 - 490 - 500 - 510 - 520 - 530 - 540 - 550 - 560 - 570 - 580 - 590 - 600 - 610 - 620 - 630 - 640 - 650 - 660 - 670 - 680 - 690 - 700 - 710 - 720 - 730 - 740 - 750 - 760 - 770 - 780 - 790 - 800 - 810 - 820 - 830 - 840 - 850 - 860 - 870 - 880 - 890 - 900 - 910 - 920 - 930 - 940 - 950 - 960 - 970 - 980 - 990 - 1000

Notes: ____

____

____

______________________________ **End Date:** ____________

Kick: ______________________________ **Start Date:** __________

Kicking Leg: Left - Right - Alt. **Kicking Height: Low - Mid - High**

Notes: ____

____

____

Slow Reps: 10 - 20 - 30 - 40 - 50 - 60 - 70 - 80 - 90 - 100 - 110 - 120 - 130
140 - 150 - 160 - 170 - 180 - 190 - 200 - 210 - 220 - 230 - 240 - 250 - 260
270 - 280 - 290 - 300 - 310 - 320 - 330 - 340 - 350 - 360 - 370 - 380 - 390
400 - 410 - 420 - 430 - 440 - 450 - 460 - 470 - 480 - 490 - 500 - 510 - 520
530 - 540 - 550 - 560 - 570 - 580 - 590 - 600 - 610 - 620 - 630 - 640 - 650
660 - 670 - 680 - 690 - 700 - 710 - 720 - 730 - 740 - 750 - 760 - 770 - 780
790 - 800 - 810 - 820 - 830 - 840 - 850 - 860 - 870 - 880 - 890 - 900 - 910
920 - 930 - 940 - 950 - 960 - 970 - 980 - 990 - 1000

Reg Reps: 10 - 20 - 30 - 40 - 50 - 60 - 70 - 80 - 90 - 100 - 110 - 120 - 130
140 - 150 - 160 - 170 - 180 - 190 - 200 - 210 - 220 - 230 - 240 - 250 - 260
270 - 280 - 290 - 300 - 310 - 320 - 330 - 340 - 350 - 360 - 370 - 380 - 390
400 - 410 - 420 - 430 - 440 - 450 - 460 - 470 - 480 - 490 - 500 - 510 - 520
530 - 540 - 550 - 560 - 570 - 580 - 590 - 600 - 610 - 620 - 630 - 640 - 650
660 - 670 - 680 - 690 - 700 - 710 - 720 - 730 - 740 - 750 - 760 - 770 - 780
790 - 800 - 810 - 820 - 830 - 840 - 850 - 860 - 870 - 880 - 890 - 900 - 910
920 - 930 - 940 - 950 - 960 - 970 - 980 - 990 - 1000

Notes: ____

____

____

____

Reg Reps: 10 - 20 - 30 - 40 - 50 - 60 - 70 - 80 - 90 - 100 - 110 - 120 - 130
140 - 150 - 160 - 170 - 180 - 190 - 200 - 210 - 220 - 230 - 240 - 250 - 260
270 - 280 - 290 - 300 - 310 - 320 - 330 - 340 - 350 - 360 - 370 - 380 - 390
400 - 410 - 420 - 430 - 440 - 450 - 460 - 470 - 480 - 490 - 500 - 510 - 520
530 - 540 - 550 - 560 - 570 - 580 - 590 - 600 - 610 - 620 - 630 - 640 - 650
660 - 670 - 680 - 690 - 700 - 710 - 720 - 730 - 740 - 750 - 760 - 770 - 780
790 - 800 - 810 - 820 - 830 - 840 - 850 - 860 - 870 - 880 - 890 - 900 - 910
920 - 930 - 940 - 950 - 960 - 970 - 980 - 990 - 1000

Notes: ____

____

____

____

Reg Reps: 10 - 20 - 30 - 40 - 50 - 60 - 70 - 80 - 90 - 100 - 110 - 120 - 130
140 - 150 - 160 - 170 - 180 - 190 - 200 - 210 - 220 - 230 - 240 - 250 - 260
270 - 280 - 290 - 300 - 310 - 320 - 330 - 340 - 350 - 360 - 370 - 380 - 390
400 - 410 - 420 - 430 - 440 - 450 - 460 - 470 - 480 - 490 - 500 - 510 - 520
530 - 540 - 550 - 560 - 570 - 580 - 590 - 600 - 610 - 620 - 630 - 640 - 650
660 - 670 - 680 - 690 - 700 - 710 - 720 - 730 - 740 - 750 - 760 - 770 - 780
790 - 800 - 810 - 820 - 830 - 840 - 850 - 860 - 870 - 880 - 890 - 900 - 910
920 - 930 - 940 - 950 - 960 - 970 - 980 - 990 - 1000

Notes: ____

____

____

____

Reg Reps: 10 - 20 - 30 - 40 - 50 - 60 - 70 - 80 - 90 - 100 - 110 - 120 - 130
140 - 150 - 160 - 170 - 180 - 190 - 200 - 210 - 220 - 230 - 240 - 250 - 260
270 - 280 - 290 - 300 - 310 - 320 - 330 - 340 - 350 - 360 - 370 - 380 - 390
400 - 410 - 420 - 430 - 440 - 450 - 460 - 470 - 480 - 490 - 500 - 510 - 520
530 - 540 - 550 - 560 - 570 - 580 - 590 - 600 - 610 - 620 - 630 - 640 - 650
660 - 670 - 680 - 690 - 700 - 710 - 720 - 730 - 740 - 750 - 760 - 770 - 780
790 - 800 - 810 - 820 - 830 - 840 - 850 - 860 - 870 - 880 - 890 - 900 - 910
920 - 930 - 940 - 950 - 960 - 970 - 980 - 990 - 1000

Notes: ____

____

____

____

Reg Reps: 10 - 20 - 30 - 40 - 50 - 60 - 70 - 80 - 90 - 100 - 110 - 120 - 130
140 - 150 - 160 - 170 - 180 - 190 - 200 - 210 - 220 - 230 - 240 - 250 - 260
270 - 280 - 290 - 300 - 310 - 320 - 330 - 340 - 350 - 360 - 370 - 380 - 390
400 - 410 - 420 - 430 - 440 - 450 - 460 - 470 - 480 - 490 - 500 - 510 - 520
530 - 540 - 550 - 560 - 570 - 580 - 590 - 600 - 610 - 620 - 630 - 640 - 650
660 - 670 - 680 - 690 - 700 - 710 - 720 - 730 - 740 - 750 - 760 - 770 - 780
790 - 800 - 810 - 820 - 830 - 840 - 850 - 860 - 870 - 880 - 890 - 900 - 910
920 - 930 - 940 - 950 - 960 - 970 - 980 - 990 - 1000

Notes: ____

____

____

______________________________**End Date:**__________

Kick:________________________________ **Start Date:** _________

Kicking Leg: Left - Right - Alt. **Kicking Height: Low - Mid - High**

Notes:____

Slow Reps: 10 - 20 - 30 - 40 - 50 - 60 - 70 - 80 - 90 - 100 - 110 - 120 - 130
140 - 150 - 160 - 170 - 180 - 190 - 200 - 210 - 220 - 230 - 240 - 250 - 260
270 - 280 - 290 - 300 - 310 - 320 - 330 - 340 - 350 - 360 - 370 - 380 - 390
400 - 410 - 420 - 430 - 440 - 450 - 460 - 470 - 480 - 490 - 500 - 510 - 520
530 - 540 - 550 - 560 - 570 - 580 - 590 - 600 - 610 - 620 - 630 - 640 - 650
660 - 670 - 680 - 690 - 700 - 710 - 720 - 730 - 740 - 750 - 760 - 770 - 780
790 - 800 - 810 - 820 - 830 - 840 - 850 - 860 - 870 - 880 - 890 - 900 - 910
920 - 930 - 940 - 950 - 960 - 970 - 980 - 990 - 1000

Reg Reps: 10 - 20 - 30 - 40 - 50 - 60 - 70 - 80 - 90 - 100 - 110 - 120 - 130
140 - 150 - 160 - 170 - 180 - 190 - 200 - 210 - 220 - 230 - 240 - 250 - 260
270 - 280 - 290 - 300 - 310 - 320 - 330 - 340 - 350 - 360 - 370 - 380 - 390
400 - 410 - 420 - 430 - 440 - 450 - 460 - 470 - 480 - 490 - 500 - 510 - 520
530 - 540 - 550 - 560 - 570 - 580 - 590 - 600 - 610 - 620 - 630 - 640 - 650
660 - 670 - 680 - 690 - 700 - 710 - 720 - 730 - 740 - 750 - 760 - 770 - 780
790 - 800 - 810 - 820 - 830 - 840 - 850 - 860 - 870 - 880 - 890 - 900 - 910
920 - 930 - 940 - 950 - 960 - 970 - 980 - 990 - 1000

Notes: ____

Reg Reps: 10 - 20 - 30 - 40 - 50 - 60 - 70 - 80 - 90 - 100 - 110 - 120 - 130
140 - 150 - 160 - 170 - 180 - 190 - 200 - 210 - 220 - 230 - 240 - 250 - 260
270 - 280 - 290 - 300 - 310 - 320 - 330 - 340 - 350 - 360 - 370 - 380 - 390
400 - 410 - 420 - 430 - 440 - 450 - 460 - 470 - 480 - 490 - 500 - 510 - 520
530 - 540 - 550 - 560 - 570 - 580 - 590 - 600 - 610 - 620 - 630 - 640 - 650
660 - 670 - 680 - 690 - 700 - 710 - 720 - 730 - 740 - 750 - 760 - 770 - 780
790 - 800 - 810 - 820 - 830 - 840 - 850 - 860 - 870 - 880 - 890 - 900 - 910
920 - 930 - 940 - 950 - 960 - 970 - 980 - 990 - 1000

Notes: ____

Reg Reps: 10 - 20 - 30 - 40 - 50 - 60 - 70 - 80 - 90 - 100 - 110 - 120 - 130
140 - 150 - 160 - 170 - 180 - 190 - 200 - 210 - 220 - 230 - 240 - 250 - 260
270 - 280 - 290 - 300 - 310 - 320 - 330 - 340 - 350 - 360 - 370 - 380 - 390
400 - 410 - 420 - 430 - 440 - 450 - 460 - 470 - 480 - 490 - 500 - 510 - 520
530 - 540 - 550 - 560 - 570 - 580 - 590 - 600 - 610 - 620 - 630 - 640 - 650
660 - 670 - 680 - 690 - 700 - 710 - 720 - 730 - 740 - 750 - 760 - 770 - 780
790 - 800 - 810 - 820 - 830 - 840 - 850 - 860 - 870 - 880 - 890 - 900 - 910
920 - 930 - 940 - 950 - 960 - 970 - 980 - 990 - 1000

Notes: ____

____

____

____

Reg Reps: 10 - 20 - 30 - 40 - 50 - 60 - 70 - 80 - 90 - 100 - 110 - 120 - 130
140 - 150 - 160 - 170 - 180 - 190 - 200 - 210 - 220 - 230 - 240 - 250 - 260
270 - 280 - 290 - 300 - 310 - 320 - 330 - 340 - 350 - 360 - 370 - 380 - 390
400 - 410 - 420 - 430 - 440 - 450 - 460 - 470 - 480 - 490 - 500 - 510 - 520
530 - 540 - 550 - 560 - 570 - 580 - 590 - 600 - 610 - 620 - 630 - 640 - 650
660 - 670 - 680 - 690 - 700 - 710 - 720 - 730 - 740 - 750 - 760 - 770 - 780
790 - 800 - 810 - 820 - 830 - 840 - 850 - 860 - 870 - 880 - 890 - 900 - 910
920 - 930 - 940 - 950 - 960 - 970 - 980 - 990 - 1000

Notes: ____

____

____

____

Reg Reps: 10 - 20 - 30 - 40 - 50 - 60 - 70 - 80 - 90 - 100 - 110 - 120 - 130
140 - 150 - 160 - 170 - 180 - 190 - 200 - 210 - 220 - 230 - 240 - 250 - 260
270 - 280 - 290 - 300 - 310 - 320 - 330 - 340 - 350 - 360 - 370 - 380 - 390
400 - 410 - 420 - 430 - 440 - 450 - 460 - 470 - 480 - 490 - 500 - 510 - 520
530 - 540 - 550 - 560 - 570 - 580 - 590 - 600 - 610 - 620 - 630 - 640 - 650
660 - 670 - 680 - 690 - 700 - 710 - 720 - 730 - 740 - 750 - 760 - 770 - 780
790 - 800 - 810 - 820 - 830 - 840 - 850 - 860 - 870 - 880 - 890 - 900 - 910
920 - 930 - 940 - 950 - 960 - 970 - 980 - 990 - 1000

Notes: ____

____

____

______________________________ **End Date:** ____________

Kick: ______________________________ **Start Date:** ________

Kicking Leg: Left - Right - Alt. **Kicking Height: Low - Mid - High**

Notes: ____

____

____

Slow Reps: 10 - 20 - 30 - 40 - 50 - 60 - 70 - 80 - 90 - 100 - 110 - 120 - 130
140 - 150 - 160 - 170 - 180 - 190 - 200 - 210 - 220 - 230 - 240 - 250 - 260
270 - 280 - 290 - 300 - 310 - 320 - 330 - 340 - 350 - 360 - 370 - 380 - 390
400 - 410 - 420 - 430 - 440 - 450 - 460 - 470 - 480 - 490 - 500 - 510 - 520
530 - 540 - 550 - 560 - 570 - 580 - 590 - 600 - 610 - 620 - 630 - 640 - 650
660 - 670 - 680 - 690 - 700 - 710 - 720 - 730 - 740 - 750 - 760 - 770 - 780
790 - 800 - 810 - 820 - 830 - 840 - 850 - 860 - 870 - 880 - 890 - 900 - 910
920 - 930 - 940 - 950 - 960 - 970 - 980 - 990 - 1000

Reg Reps: 10 - 20 - 30 - 40 - 50 - 60 - 70 - 80 - 90 - 100 - 110 - 120 - 130
140 - 150 - 160 - 170 - 180 - 190 - 200 - 210 - 220 - 230 - 240 - 250 - 260
270 - 280 - 290 - 300 - 310 - 320 - 330 - 340 - 350 - 360 - 370 - 380 - 390
400 - 410 - 420 - 430 - 440 - 450 - 460 - 470 - 480 - 490 - 500 - 510 - 520
530 - 540 - 550 - 560 - 570 - 580 - 590 - 600 - 610 - 620 - 630 - 640 - 650
660 - 670 - 680 - 690 - 700 - 710 - 720 - 730 - 740 - 750 - 760 - 770 - 780
790 - 800 - 810 - 820 - 830 - 840 - 850 - 860 - 870 - 880 - 890 - 900 - 910
920 - 930 - 940 - 950 - 960 - 970 - 980 - 990 - 1000

Notes: ____

____

____

____

Reg Reps: 10 - 20 - 30 - 40 - 50 - 60 - 70 - 80 - 90 - 100 - 110 - 120 - 130
140 - 150 - 160 - 170 - 180 - 190 - 200 - 210 - 220 - 230 - 240 - 250 - 260
270 - 280 - 290 - 300 - 310 - 320 - 330 - 340 - 350 - 360 - 370 - 380 - 390
400 - 410 - 420 - 430 - 440 - 450 - 460 - 470 - 480 - 490 - 500 - 510 - 520
530 - 540 - 550 - 560 - 570 - 580 - 590 - 600 - 610 - 620 - 630 - 640 - 650
660 - 670 - 680 - 690 - 700 - 710 - 720 - 730 - 740 - 750 - 760 - 770 - 780
790 - 800 - 810 - 820 - 830 - 840 - 850 - 860 - 870 - 880 - 890 - 900 - 910
920 - 930 - 940 - 950 - 960 - 970 - 980 - 990 - 1000

Notes: ____

____

____

____

Reg Reps: 10 - 20 - 30 - 40 - 50 - 60 - 70 - 80 - 90 - 100 - 110 - 120 - 130 - 140 - 150 - 160 - 170 - 180 - 190 - 200 - 210 - 220 - 230 - 240 - 250 - 260 - 270 - 280 - 290 - 300 - 310 - 320 - 330 - 340 - 350 - 360 - 370 - 380 - 390 - 400 - 410 - 420 - 430 - 440 - 450 - 460 - 470 - 480 - 490 - 500 - 510 - 520 - 530 - 540 - 550 - 560 - 570 - 580 - 590 - 600 - 610 - 620 - 630 - 640 - 650 - 660 - 670 - 680 - 690 - 700 - 710 - 720 - 730 - 740 - 750 - 760 - 770 - 780 - 790 - 800 - 810 - 820 - 830 - 840 - 850 - 860 - 870 - 880 - 890 - 900 - 910 - 920 - 930 - 940 - 950 - 960 - 970 - 980 - 990 - 1000

Notes: ____

____

____

____

Reg Reps: 10 - 20 - 30 - 40 - 50 - 60 - 70 - 80 - 90 - 100 - 110 - 120 - 130 - 140 - 150 - 160 - 170 - 180 - 190 - 200 - 210 - 220 - 230 - 240 - 250 - 260 - 270 - 280 - 290 - 300 - 310 - 320 - 330 - 340 - 350 - 360 - 370 - 380 - 390 - 400 - 410 - 420 - 430 - 440 - 450 - 460 - 470 - 480 - 490 - 500 - 510 - 520 - 530 - 540 - 550 - 560 - 570 - 580 - 590 - 600 - 610 - 620 - 630 - 640 - 650 - 660 - 670 - 680 - 690 - 700 - 710 - 720 - 730 - 740 - 750 - 760 - 770 - 780 - 790 - 800 - 810 - 820 - 830 - 840 - 850 - 860 - 870 - 880 - 890 - 900 - 910 - 920 - 930 - 940 - 950 - 960 - 970 - 980 - 990 - 1000

Notes: ____

____

____

____

Reg Reps: 10 - 20 - 30 - 40 - 50 - 60 - 70 - 80 - 90 - 100 - 110 - 120 - 130 - 140 - 150 - 160 - 170 - 180 - 190 - 200 - 210 - 220 - 230 - 240 - 250 - 260 - 270 - 280 - 290 - 300 - 310 - 320 - 330 - 340 - 350 - 360 - 370 - 380 - 390 - 400 - 410 - 420 - 430 - 440 - 450 - 460 - 470 - 480 - 490 - 500 - 510 - 520 - 530 - 540 - 550 - 560 - 570 - 580 - 590 - 600 - 610 - 620 - 630 - 640 - 650 - 660 - 670 - 680 - 690 - 700 - 710 - 720 - 730 - 740 - 750 - 760 - 770 - 780 - 790 - 800 - 810 - 820 - 830 - 840 - 850 - 860 - 870 - 880 - 890 - 900 - 910 - 920 - 930 - 940 - 950 - 960 - 970 - 980 - 990 - 1000

Notes: ____

____

____

______________________________ **End Date:** ____________

Kick:__________________________________ **Start Date:** _________

Kicking Leg: Left - Right - Alt. **Kicking Height: Low - Mid - High**

Notes:____

Slow Reps: 10 - 20 - 30 - 40 - 50 - 60 - 70 - 80 - 90 - 100 - 110 - 120 - 130
140 - 150 - 160 - 170 - 180 - 190 - 200 - 210 - 220 - 230 - 240 - 250 - 260
270 - 280 - 290 - 300 - 310 - 320 - 330 - 340 - 350 - 360 - 370 - 380 - 390
400 - 410 - 420 - 430 - 440 - 450 - 460 - 470 - 480 - 490 - 500 - 510 - 520
530 - 540 - 550 - 560 - 570 - 580 - 590 - 600 - 610 - 620 - 630 - 640 - 650
660 - 670 - 680 - 690 - 700 - 710 - 720 - 730 - 740 - 750 - 760 - 770 - 780
790 - 800 - 810 - 820 - 830 - 840 - 850 - 860 - 870 - 880 - 890 - 900 - 910
920 - 930 - 940 - 950 - 960 - 970 - 980 - 990 - 1000

Reg Reps: 10 - 20 - 30 - 40 - 50 - 60 - 70 - 80 - 90 - 100 - 110 - 120 - 130
140 - 150 - 160 - 170 - 180 - 190 - 200 - 210 - 220 - 230 - 240 - 250 - 260
270 - 280 - 290 - 300 - 310 - 320 - 330 - 340 - 350 - 360 - 370 - 380 - 390
400 - 410 - 420 - 430 - 440 - 450 - 460 - 470 - 480 - 490 - 500 - 510 - 520
530 - 540 - 550 - 560 - 570 - 580 - 590 - 600 - 610 - 620 - 630 - 640 - 650
660 - 670 - 680 - 690 - 700 - 710 - 720 - 730 - 740 - 750 - 760 - 770 - 780
790 - 800 - 810 - 820 - 830 - 840 - 850 - 860 - 870 - 880 - 890 - 900 - 910
920 - 930 - 940 - 950 - 960 - 970 - 980 - 990 - 1000

Notes: ____

Reg Reps: 10 - 20 - 30 - 40 - 50 - 60 - 70 - 80 - 90 - 100 - 110 - 120 - 130
140 - 150 - 160 - 170 - 180 - 190 - 200 - 210 - 220 - 230 - 240 - 250 - 260
270 - 280 - 290 - 300 - 310 - 320 - 330 - 340 - 350 - 360 - 370 - 380 - 390
400 - 410 - 420 - 430 - 440 - 450 - 460 - 470 - 480 - 490 - 500 - 510 - 520
530 - 540 - 550 - 560 - 570 - 580 - 590 - 600 - 610 - 620 - 630 - 640 - 650
660 - 670 - 680 - 690 - 700 - 710 - 720 - 730 - 740 - 750 - 760 - 770 - 780
790 - 800 - 810 - 820 - 830 - 840 - 850 - 860 - 870 - 880 - 890 - 900 - 910
920 - 930 - 940 - 950 - 960 - 970 - 980 - 990 - 1000

Notes: ____

Reg Reps: 10 - 20 - 30 - 40 - 50 - 60 - 70 - 80 - 90 - 100 - 110 - 120 - 130
140 - 150 - 160 - 170 - 180 - 190 - 200 - 210 - 220 - 230 - 240 - 250 - 260
270 - 280 - 290 - 300 - 310 - 320 - 330 - 340 - 350 - 360 - 370 - 380 - 390
400 - 410 - 420 - 430 - 440 - 450 - 460 - 470 - 480 - 490 - 500 - 510 - 520
530 - 540 - 550 - 560 - 570 - 580 - 590 - 600 - 610 - 620 - 630 - 640 - 650
660 - 670 - 680 - 690 - 700 - 710 - 720 - 730 - 740 - 750 - 760 - 770 - 780
790 - 800 - 810 - 820 - 830 - 840 - 850 - 860 - 870 - 880 - 890 - 900 - 910
920 - 930 - 940 - 950 - 960 - 970 - 980 - 990 - 1000

Notes: ____

____

____

____

Reg Reps: 10 - 20 - 30 - 40 - 50 - 60 - 70 - 80 - 90 - 100 - 110 - 120 - 130
140 - 150 - 160 - 170 - 180 - 190 - 200 - 210 - 220 - 230 - 240 - 250 - 260
270 - 280 - 290 - 300 - 310 - 320 - 330 - 340 - 350 - 360 - 370 - 380 - 390
400 - 410 - 420 - 430 - 440 - 450 - 460 - 470 - 480 - 490 - 500 - 510 - 520
530 - 540 - 550 - 560 - 570 - 580 - 590 - 600 - 610 - 620 - 630 - 640 - 650
660 - 670 - 680 - 690 - 700 - 710 - 720 - 730 - 740 - 750 - 760 - 770 - 780
790 - 800 - 810 - 820 - 830 - 840 - 850 - 860 - 870 - 880 - 890 - 900 - 910
920 - 930 - 940 - 950 - 960 - 970 - 980 - 990 - 1000

Notes: ____

____

____

____

Reg Reps: 10 - 20 - 30 - 40 - 50 - 60 - 70 - 80 - 90 - 100 - 110 - 120 - 130
140 - 150 - 160 - 170 - 180 - 190 - 200 - 210 - 220 - 230 - 240 - 250 - 260
270 - 280 - 290 - 300 - 310 - 320 - 330 - 340 - 350 - 360 - 370 - 380 - 390
400 - 410 - 420 - 430 - 440 - 450 - 460 - 470 - 480 - 490 - 500 - 510 - 520
530 - 540 - 550 - 560 - 570 - 580 - 590 - 600 - 610 - 620 - 630 - 640 - 650
660 - 670 - 680 - 690 - 700 - 710 - 720 - 730 - 740 - 750 - 760 - 770 - 780
790 - 800 - 810 - 820 - 830 - 840 - 850 - 860 - 870 - 880 - 890 - 900 - 910
920 - 930 - 940 - 950 - 960 - 970 - 980 - 990 - 1000

Notes: ____

____

____

______________________________ **End Date:** ____________

Kick:____**Start Date:** __________
Kicking Leg: Left - Right - Alt. **Kicking Height: Low - Mid - High**
Notes:____
____
____

Slow Reps: 10 - 20 - 30 - 40 - 50 - 60 - 70 - 80 - 90 - 100 - 110 - 120 - 130
140 - 150 - 160 - 170 - 180 - 190 - 200 - 210 - 220 - 230 - 240 - 250 - 260
270 - 280 - 290 - 300 - 310 - 320 - 330 - 340 - 350 - 360 - 370 - 380 - 390
400 - 410 - 420 - 430 - 440 - 450 - 460 - 470 - 480 - 490 - 500 - 510 - 520
530 - 540 - 550 - 560 - 570 - 580 - 590 - 600 - 610 - 620 - 630 - 640 - 650
660 - 670 - 680 - 690 - 700 - 710 - 720 - 730 - 740 - 750 - 760 - 770 - 780
790 - 800 - 810 - 820 - 830 - 840 - 850 - 860 - 870 - 880 - 890 - 900 - 910
920 - 930 - 940 - 950 - 960 - 970 - 980 - 990 - 1000

Reg Reps: 10 - 20 - 30 - 40 - 50 - 60 - 70 - 80 - 90 - 100 - 110 - 120 - 130
140 - 150 - 160 - 170 - 180 - 190 - 200 - 210 - 220 - 230 - 240 - 250 - 260
270 - 280 - 290 - 300 - 310 - 320 - 330 - 340 - 350 - 360 - 370 - 380 - 390
400 - 410 - 420 - 430 - 440 - 450 - 460 - 470 - 480 - 490 - 500 - 510 - 520
530 - 540 - 550 - 560 - 570 - 580 - 590 - 600 - 610 - 620 - 630 - 640 - 650
660 - 670 - 680 - 690 - 700 - 710 - 720 - 730 - 740 - 750 - 760 - 770 - 780
790 - 800 - 810 - 820 - 830 - 840 - 850 - 860 - 870 - 880 - 890 - 900 - 910
920 - 930 - 940 - 950 - 960 - 970 - 980 - 990 - 1000
Notes: ____
____
____
____

Reg Reps: 10 - 20 - 30 - 40 - 50 - 60 - 70 - 80 - 90 - 100 - 110 - 120 - 130
140 - 150 - 160 - 170 - 180 - 190 - 200 - 210 - 220 - 230 - 240 - 250 - 260
270 - 280 - 290 - 300 - 310 - 320 - 330 - 340 - 350 - 360 - 370 - 380 - 390
400 - 410 - 420 - 430 - 440 - 450 - 460 - 470 - 480 - 490 - 500 - 510 - 520
530 - 540 - 550 - 560 - 570 - 580 - 590 - 600 - 610 - 620 - 630 - 640 - 650
660 - 670 - 680 - 690 - 700 - 710 - 720 - 730 - 740 - 750 - 760 - 770 - 780
790 - 800 - 810 - 820 - 830 - 840 - 850 - 860 - 870 - 880 - 890 - 900 - 910
920 - 930 - 940 - 950 - 960 - 970 - 980 - 990 - 1000
Notes: ____
____
____
____

Reg Reps: 10 - 20 - 30 - 40 - 50 - 60 - 70 - 80 - 90 - 100 - 110 - 120 - 130 - 140 - 150 - 160 - 170 - 180 - 190 - 200 - 210 - 220 - 230 - 240 - 250 - 260 - 270 - 280 - 290 - 300 - 310 - 320 - 330 - 340 - 350 - 360 - 370 - 380 - 390 - 400 - 410 - 420 - 430 - 440 - 450 - 460 - 470 - 480 - 490 - 500 - 510 - 520 - 530 - 540 - 550 - 560 - 570 - 580 - 590 - 600 - 610 - 620 - 630 - 640 - 650 - 660 - 670 - 680 - 690 - 700 - 710 - 720 - 730 - 740 - 750 - 760 - 770 - 780 - 790 - 800 - 810 - 820 - 830 - 840 - 850 - 860 - 870 - 880 - 890 - 900 - 910 - 920 - 930 - 940 - 950 - 960 - 970 - 980 - 990 - 1000

Notes: ______________________________

____

____

____

Reg Reps: 10 - 20 - 30 - 40 - 50 - 60 - 70 - 80 - 90 - 100 - 110 - 120 - 130 - 140 - 150 - 160 - 170 - 180 - 190 - 200 - 210 - 220 - 230 - 240 - 250 - 260 - 270 - 280 - 290 - 300 - 310 - 320 - 330 - 340 - 350 - 360 - 370 - 380 - 390 - 400 - 410 - 420 - 430 - 440 - 450 - 460 - 470 - 480 - 490 - 500 - 510 - 520 - 530 - 540 - 550 - 560 - 570 - 580 - 590 - 600 - 610 - 620 - 630 - 640 - 650 - 660 - 670 - 680 - 690 - 700 - 710 - 720 - 730 - 740 - 750 - 760 - 770 - 780 - 790 - 800 - 810 - 820 - 830 - 840 - 850 - 860 - 870 - 880 - 890 - 900 - 910 - 920 - 930 - 940 - 950 - 960 - 970 - 980 - 990 - 1000

Notes: ______________________________

____

____

____

Reg Reps: 10 - 20 - 30 - 40 - 50 - 60 - 70 - 80 - 90 - 100 - 110 - 120 - 130 - 140 - 150 - 160 - 170 - 180 - 190 - 200 - 210 - 220 - 230 - 240 - 250 - 260 - 270 - 280 - 290 - 300 - 310 - 320 - 330 - 340 - 350 - 360 - 370 - 380 - 390 - 400 - 410 - 420 - 430 - 440 - 450 - 460 - 470 - 480 - 490 - 500 - 510 - 520 - 530 - 540 - 550 - 560 - 570 - 580 - 590 - 600 - 610 - 620 - 630 - 640 - 650 - 660 - 670 - 680 - 690 - 700 - 710 - 720 - 730 - 740 - 750 - 760 - 770 - 780 - 790 - 800 - 810 - 820 - 830 - 840 - 850 - 860 - 870 - 880 - 890 - 900 - 910 - 920 - 930 - 940 - 950 - 960 - 970 - 980 - 990 - 1000

Notes: ______________________________

____

____

______________________________ **End Date:** __________

ADDITIONAL NOTES

Previous Page:_______ Kick:________________________________ Rep #:_____

Previous Page:_______ Kick:________________________________ Rep #:_____

Previous Page:_______ Kick:________________________________ Rep #:_____

Previous Page:_______ Kick:________________________________ Rep #:_______

ADDITIONAL NOTES

Previous Page:_______ Kick:________________________________ Rep #:_____

Previous Page:_______ Kick:________________________________ Rep #:_____

Previous Page:_______ Kick:________________________________ Rep #:_____

Previous Page:_______ Kick:________________________________ Rep #:_____

ADDITIONAL NOTES

Previous Page:_______ **Kick:**________________________________ **Rep #:**_____

Previous Page:_______ **Kick:**________________________________ **Rep #:**_____

Previous Page:_______ **Kick:**________________________________ **Rep #:**_____

Previous Page:_______ **Kick:**________________________________ **Rep #:**_____

ADDITIONAL NOTES

Previous Page:_______ Kick:_______________________________ Rep #:_____

Previous Page:_______ Kick:_______________________________ Rep #:_____

Previous Page:_______ Kick:_______________________________ Rep #:_____

Previous Page:_______ Kick:_______________________________ Rep #:_____

ADDITIONAL NOTES

Previous Page:_______ **Kick:**________________________________ **Rep #:**_____

Previous Page:_______ **Kick:**________________________________ **Rep #:**_____

Previous Page:_______ **Kick:**________________________________ **Rep #:**_____

Previous Page:_______ **Kick:**________________________________ **Rep #:**_____

ADDITIONAL NOTES

Previous Page:_______ Kick:______________________________ Rep #:_____

Previous Page:_______ Kick:______________________________ Rep #:_____

Previous Page:_______ Kick:______________________________ Rep #:_____

Previous Page:_______ Kick:______________________________ Rep #:_____

ADDITIONAL NOTES

Previous Page:_______ Kick:________________________________ Rep #:_____

Previous Page:_______ Kick:________________________________ Rep #:_____

Previous Page:_______ Kick:________________________________ Rep #:_____

Previous Page:_______ Kick:________________________________ Rep #:_____

ADDITIONAL NOTES

Previous Page:_______ Kick:________________________________ Rep #:_____

Previous Page:_______ Kick:________________________________ Rep #:_____

Previous Page:_______ Kick:________________________________ Rep #:_____

Previous Page:_______ Kick:________________________________ Rep #:_______

ADDITIONAL NOTES

Previous Page:_______ Kick:________________________________ Rep #:_____

Previous Page:_______ Kick:________________________________ Rep #:_____

Previous Page:_______ Kick:________________________________ Rep #:_____

Previous Page:_______ Kick:________________________________ Rep #:_______

ADDITIONAL NOTES

Previous Page:_______ Kick:_______________________________ Rep #:_____

Previous Page:_______ Kick:_______________________________ Rep #:_____

Previous Page:_______ Kick:_______________________________ Rep #:_____

Previous Page:_______ Kick:_______________________________ Rep #:_____

ADDITIONAL NOTES

Previous Page:_______ Kick:_______________________________ Rep #:_____

Previous Page:_______ Kick:_______________________________ Rep #:_____

Previous Page:_______ Kick:_______________________________ Rep #:_____

Previous Page:_______ Kick:_______________________________ Rep #:_______

ADDITIONAL NOTES

Previous Page:_______ **Kick:**________________________________ **Rep #:**_____

Previous Page:_______ **Kick:**________________________________ **Rep #:**_____

Previous Page:_______ **Kick:**________________________________ **Rep #:**_____

Previous Page:_______ **Kick:**________________________________ **Rep #:**_____

ADDITIONAL NOTES

Previous Page:_______ Kick:________________________________ Rep #:_____

Previous Page:_______ Kick:________________________________ Rep #:_____

Previous Page:_______ Kick:________________________________ Rep #:_____

Previous Page:_______ Kick:________________________________ Rep #:_______

ADDITIONAL NOTES

Previous Page:_______ Kick:________________________________ Rep #:_____

Previous Page:_______ Kick:________________________________ Rep #:_____

Previous Page:_______ Kick:________________________________ Rep #:_____

Previous Page:_______ Kick:________________________________ Rep #:_____

ADDITIONAL NOTES

Previous Page:_______ Kick:_______________________________ Rep #:_____

Previous Page:_______ Kick:_______________________________ Rep #:_____

Previous Page:_______ Kick:_______________________________ Rep #:_____

Previous Page:_______ Kick:_______________________________ Rep #:_____

ADDITIONAL NOTES

Previous Page:_______ **Kick:**________________________________ **Rep #:**_____

Previous Page:_______ **Kick:**________________________________ **Rep #:**_____

Previous Page:_______ **Kick:**________________________________ **Rep #:**_____

Previous Page:_______ **Kick:**________________________________ **Rep #:**_____

Made in the USA
Monee, IL
04 November 2026

67377447R00075